CARLSON

POCKET
HANDBOOK FOR
SOLID-LIQUID
SEPARATIONS

Gulf Publishing Company
Book Division
Houston, London, Paris, Tokyo

POCKET
HANDBOOK FOR
SOLID-LIQUID
SEPARATIONS

Nicholas P. Cheremisinoff

Pocket Handbook for Solid-Liquid Separations

Library of Congress Cataloging in Publication Data

Cheremisinoff, Nicholas P.
    Pocket handbook for solid-liquid separations.

    Includes index.
    1. Separation (Technology)—Handbooks, manuals, etc.
I. Title.
TP156.S45C47 1984   660.2′842   84-8970
ISBN 0-87201-830-X

CONTENTS

4.

5.

6.

7.

8.

PREFACE

Solid-liquid separation is widely practiced throughout the process and allied industries, as well as in wastewater treatment. Because of the diversified applications, and the varying properties of suspensions and sludges, equipment design and selection are often confusing and highly empirical. This pocket manual is prepared as a working guide to the principles and methods of solid-liquid separations. It is a compilation of useful formulas and scale-up criteria for commonly used separations equipment, as well as an introductory reference source of standard unit operations for solid-liquid separations. Each section is structured for quick reference through compiled short notes. Chemical, civil, and hydraulic engineers, as well as students, will find this to be a handy manual.

Nicholas P. Cheremisinoff

NOTATION

A	area
Ar	Archimedes number
a	contraction loss coefficient; acceleration in Chapter 7
a′	filtration coefficient
C	filtration constant
C_D	drag coefficient
C_F	fabric resistance coefficient
C_L	discharge coefficient
CCD	continuous countercurrent decanting
c	concentration
D, d	diameter
D_P	particle diameter
D_s	shaft diameter
E	ratio of mean velocity through microscreen to initial velocity
E_o	energy or power
F	friction or head loss
F_f	friction force
F_g	gravitational force
f	friction factor
g	acceleration due to gravity
H	head
h_c	cake height
h_ℓ	head loss
I	filterability index
ISV	initial settling velocity
J	inertia moment
K	resistance coefficient; Darcy's constant in Chapter 3
K′	hydraulic conductivity
K_c	filtration constant
K_e	head loss coefficient

K_s	separation number
k	coefficient
k'	intrinsic permeability
L	length
L_c	channel length
L_e	equivalent length
ℓ	height of bowl
M_H	mass
MLSS	mixed liquor suspended solids
m	weight ratio of wet to dry cake, or mass
N	horsepower
N_c	centrifugal number
n	number of rotations per unit time
P	pressure
Q	volumetric flowrate
q	seepage velocity
R	drum rotational speed
R_c	filter cake resistance
R_f	filter resistance
R_H	hydraulic radius
Re	Reynolds number
RCF	relative centrifugal force
r	radius
r_h	hydraulic radius
r_o	specific volumetric cake resistance
r_w	specific mass cake resistance
S	surface of solids per unit bed volume; mass fraction of solids in sludge in Chapter 3
S_c	mass fractions of solids in cake
SS	suspended solids
SVI	sludge volume index
s	cake compression coefficient
t	time
UC	uniformity coefficient
u	filtration rate or superficial velocity
u_r	peripheral velocity at radius r
u_s	settling velocity
V_H	volume
V_p	particle volume
V_∞	velocity of approach
V	filtrate volume
V_s	settling velocity

v	average fluid velocity
W	mass flow rate
w	velocity
x_o	weight solids in cake per unit filtrate volume
Z	elevation
Z'	number of fillets
Z_c	height of compressed sludge zone
Z_o	initial slurry concentration

Greek Symbols

α	specific flow resistance; flow regime constant in Chapter 1
β	parameter in Equation 6-13; angle in Chapter 7
γ	specific weight
ϵ	void fraction
η	concentration
γ_b	friction coefficient of shaft
γ_f	friction coefficient
μ	viscosity
ν	filtrate volume per unit filter medium area
ρ	density
σ	ratio of particle surface area to volume; density ratio in Chapter 3
τ	time
ψ	head loss coefficient
$\phi(\epsilon)$	void fraction function
ϕ	decimal fraction of screen area submerged
ω	rotational velocity

1

REVIEW OF FLUID FLOW PRINCIPLES

Estimating Frictional Pressure Losses

This subsection outlines methods for estimating pressure losses through piping and pipe components.

The pressure drop in a horizontal straight length of pipe of constant diameter can be calculated from the Fanning friction equation. Fanning friction factor f is a function of the Reynolds number and relative pipe wall roughness and is shown in Figure 1-1. For a given class of pipe material, roughness is relatively independent of pipe diameter; therefore the friction factor can be expressed as a function of Reynolds number and pipe diameter. For laminar flow (Re < 2,000), the friction factor is independent of pipe wall roughness.

The accuracy of the Fanning friction equation is $\pm 15\%$ for smooth pipe and $\pm 10\%$ for commercial steel pipe. Fouling can reduce the cross-sectional area or increase pipe wall roughness with time.

The basic equation for calculating pressure drop for liquid flow in pipes and fittings is the generalized Bernoulli equation:

$$-\frac{10^{-3}\Delta P}{\rho g} = \frac{\alpha\Delta(V^2)}{2g} + \Delta z + \frac{10^{-3}F}{g} \qquad (1\text{-}1)$$

$$\begin{array}{c} \text{Pressure} \\ \text{change} \end{array} = \begin{array}{c} \text{Kinetic} \\ \text{energy} \\ \text{change} \end{array} + \begin{array}{c} \text{Elevation} \\ \text{change} \end{array} + \begin{array}{c} \text{Friction} \\ \text{or head} \\ \text{loss} \end{array}$$

where the units in SI are as follows:

F = friction or head loss, kPa-m^3/kg
g = acceleration of gravity, 9.81 m/s

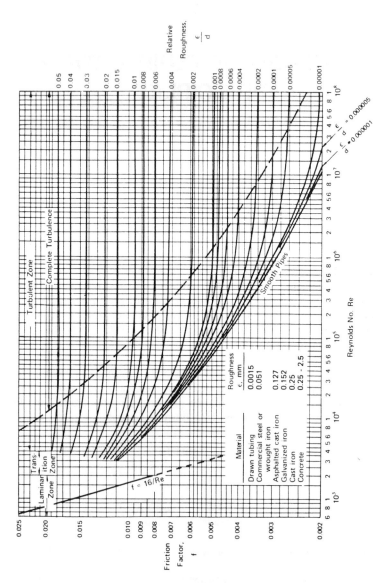

Figure 1-1. Friction factor plot for any type of commercial pipe.

ΔP = pressure change, kPa

V = velocity of the fluid, m/s

z = elevation, m

ρ = density, kg/m^3

α = constant, depending on velocity profile ($\alpha = 1.1$ for turbulent flow, $\alpha = 2.0$ for laminar flow)

For constant-diameter horizontal pipes, only the friction term on the right-hand side of the equation is important. For vertical or inclined pipes, one must include the elevation term; and for cross section changes, the kinetic energy term.

For Newtonian liquids both constant viscosity and density can be assumed. Non-Newtonian liquids are an exception to this rule. Another exception is nonisothermal flow, due either to heat exchange, or to heat production or consumption in the liquid by chemical reaction or friction losses.

Where the flow may be assumed to be isothermal across the pipe cross section, but is not isothermal along the length of the pipe, the pressure drop can be computed by dividing the pipe into a number of lengths and calculating the pressure drop in each section. Fittings that have the same nominal diameter as the pipe can be accounted for in terms of an *equivalent length* of straight pipe. This equivalent length can be computed from the resistance coefficients of the fittings. Typical values are given in Table 1-1. The equivalent length is then added to the actual length of the pipe and the sum is used in the Fanning equation for predicting the total friction pressure drop.

The pressure drop in cross section changes, such as exits and entrances of process vessels and filters, reducers, and diffusers, consists of friction and kinetic energy. Calculation of the friction loss is based on the diameter of the smaller of the two pipes with no obstructions.

For pipes ending in an area of very large cross section, such as process vessels, the frictional pressure drop is equal to the gain in pressure caused by the change in kinetic energy. The net pressure change over the cross section change is zero.

For a very gradual contraction, the friction pressure drop calculation is based on a straight piece of pipe with inside diameter equal to the narrowest cross section of the contraction.

In pressure drop calculations for lines containing fittings and cross section changes, the line is first broken into sections of constant nominal diameter. The frictional pressure drop of each change in cross section is accounted for in the equivalent length of the small-diameter pipe attached to it. The pressure drop due to the various changes in kinetic

Table 1-1
**Loss Coefficients for Turbulent Flow Through Valves
and Fittings[1]**

Type Valve Or Fitting	Head Loss Coefficient, k_e	L_e/D
Elbow, 45°	0.35	17
Elbow, 90°	0.75	35
Tee	1	50
Return Bend	1.5	75
Coupling	0.04	2
Union	0.04	2
Gate Valve—		
Wide Open	0.17	9
Half Open	4.5	225
Globe Valve—		
Wide Open	6.0	300
Half Open	9.5	475
Angle Valve—		
Wide Open	2.0	100
Check Valve—		
Ball	70.0	3,500
Swing	2.0	100
Water Meter, disk	7.0	350

energy in the line is determined by computing the overall change in kinetic energy between the inlet and outlet of the line.

When a stream is split in two or more substreams there is both a friction loss and a pressure change due to the change in kinetic energy. The same applies to the combining of streams. For tees the total pressure change is given by the equations listed in Figure 1-2.

Equations and guidelines for estimating pressure losses in piping systems are outlined as follows. Examples of single piping components are runs of straight pipe, bends, valves, and orifices. If the pipe has a noncircular cross section, first compute the equivalent hydraulic diameter:

$$d_{eq} = 4\left(\frac{\text{cross-sectional area}}{\text{wetted perimeter}}\right), \text{ in consistent units} \qquad (1\text{-}2)$$

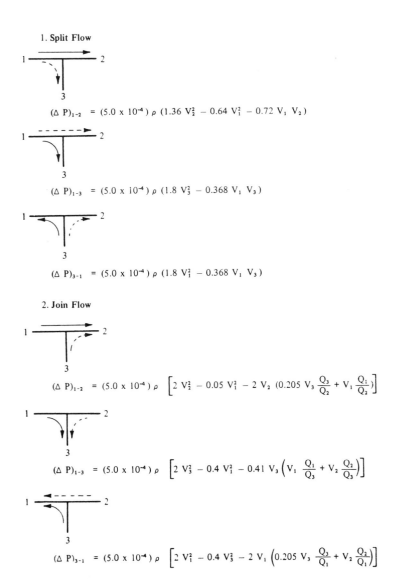

1. Split Flow

$(\Delta P)_{1-2} = (5.0 \times 10^{-4}) \rho (1.36 V_2^2 - 0.64 V_1^2 - 0.72 V_1 V_2)$

$(\Delta P)_{1-3} = (5.0 \times 10^{-4}) \rho (1.8 V_3^2 - 0.368 V_1 V_3)$

$(\Delta P)_{3-1} = (5.0 \times 10^{-4}) \rho (1.8 V_1^2 - 0.368 V_1 V_3)$

2. Join Flow

$(\Delta P)_{1-2} = (5.0 \times 10^{-4}) \rho \left[2 V_2^2 - 0.05 V_1^2 - 2 V_2 \left(0.205 V_3 \frac{Q_3}{Q_2} + V_1 \frac{Q_1}{Q_2}\right) \right]$

$(\Delta P)_{1-3} = (5.0 \times 10^{-4}) \rho \left[2 V_3^2 - 0.4 V_1^2 - 0.41 V_3 \left(V_1 \frac{Q_1}{Q_3} + V_2 \frac{Q_2}{Q_3}\right) \right]$

$(\Delta P)_{3-1} = (5.0 \times 10^{-4}) \rho \left[2 V_1^2 - 0.4 V_3^2 - 2 V_1 \left(0.205 V_3 \frac{Q_3}{Q_1} + V_2 \frac{Q_2}{Q_1}\right) \right]$

Figure 1-2. Pressure drop equations for split and join flow streams[2].

Table 1-2 gives formulas of the hydraulic radius (R_H) for various flow area geometries. The equivalent hydraulic diameter is defined as $4R_H$.

For given diameter and flow rate, compute the Reynolds number, Re, from the following equation:

$$Re = \frac{DV\rho}{\mu} = 10^{-3}\left(\frac{dV\rho}{\mu}\right)$$

$$= 1.27\left(\frac{Q\rho}{d\mu}\right) \tag{1-3}$$

$$= 1.27 \times 10^3\left(\frac{W}{d\mu}\right)$$

Table 1-2
Formulas for Hydraulic Radius for Various Flow Area Geometries

Flow Area Geometry	Hydraulic Radius, R_H	Description
Circle	$D/4$	D-diameter
Annulus	$\dfrac{(D - d)}{4}$	d-inner diameter D-outer diameter
Rectangle	$\dfrac{ab}{2(a + b)}$	a, b-sides
Triangle, Equilateral	$\dfrac{1}{12}a\sqrt{3}$	a-sides
Triangle, General	$\sqrt{\dfrac{s(s - a)(s - b)(s - c)}{(a + b + c)}}$	a, b, c-sides $s = \frac{1}{2}(a + b + c)$
Ellipse[a]	$\dfrac{ab}{k(a + b)}$	2a-major axis 2b-minor axis

[a] Values of k for z = (a − b)/(a + b); z = 0.2, 0.4, 0.6, 0.8, 1.0;
   k = 1.010, 1.040, 1.092, 1.127, 1.273

where: D = inside diameter of pipe or equivalent hydraulic diameter, m
   d = inside diameter of pipe or equivalent hydraulic diameter, cm
   Q = volumetric flow rate, dm^3/s
   Re = Reynolds number, dimensionless
   V = velocity, m/s
   W = mass flow rate, kg/s
   ρ = density, kg/m^3
   μ = dynamic viscosity, Pa - s

Figure 1-1 should then be used to obtain a value of the friction factor.
   For values of Re lower than those covered by this figure, with Re < 2,000 (Laminar Flow), calculate f from the following:

$$f = \frac{16}{Re} \tag{1-4}$$

where: f = friction factor, dimensionless.

   Next, compute the frictional pressure drop from the following equations:

$$(\Delta P)_f = 10^{-3}\left(\frac{4fL}{D}\right)\left(\frac{\rho V^2}{2}\right)$$

$$= 2\frac{fLV^2\rho}{d}$$

$$= 3.24 \times 10^6\left(\frac{fLQ^2\rho}{d^5}\right) \tag{1-5}$$

$$= 3.24 \times 10^{12}\left(\frac{fLW^2}{\rho d^5}\right)$$

where: $(\Delta P)_f$ = frictional pressure drop, kPa
   L = pipe length, m

If the pipe is not horizontal, the pressure drop due to the change in elevation must be computed:

$$(\Delta P)_e = 10^{-3}(\rho g)(z_2 - z_1) \tag{1-6}$$

where: $(\Delta P)_e$ = pressure drop due to change in elevation, kPa
z_1, z_2 = elevation of beginning and end of pipe, m

The total pressure drop is obtained by adding the frictional pressure drop $(\Delta P)_f$ and the pressure drop due to the change in elevation $(\Delta P)_e$.

Resistance coefficients for bends, ells and tees are given in Figure 1-3.

For pipes larger than 250mm ID, use the resistance coefficient for 250mm ID pipe. If the Reynolds number is such that the flow is not in the region of complete turbulence (f is constant), the value of K should be multipled by the ratio:

$$\frac{f_{\text{(at calculated Reynolds number)}}}{f_{\text{(in range of complete turbulence)}}}$$

From the resistance coefficients from Figure 1-3, the frictional pressure drop is:

$$(\Delta P)_f = 10^{-3}\left(\frac{K\rho V^2}{2}\right) \tag{1-7}$$

For long nonhorizontal bends, add the pressure drop due to the change in elevation calculated from Equation 1-6.

For blanked-off tees and Y's, use Equation 1-7 and the resistance coefficients for tees in Figure 1-3. For tees in which streams are split or joined, the pressure drop should be calculated from the equations given in Figure 1-2.

These equations account for both frictional pressure drop and pressure drop due to changes in kinetic energy. To account for entrance and exit effects in cases where the inlet leading line is short, a multiplying factor of 1.25 can be applied. Sample calculations are given by Cheremisinoff[3,4,5].

Flow Through Varying Cross Sections

During a sudden contraction, losses may be estimated from the Weisbach equation:

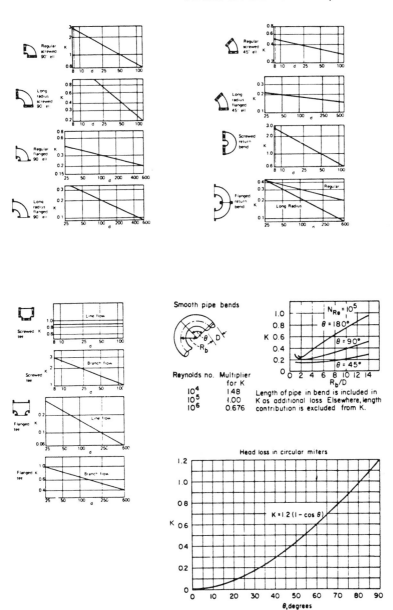

Figure 1-3. Resistance coefficients for bends, ells and tees[1].

$$h_\ell = [0.04 + (1/a - 1)^2]\frac{w_2^2}{2g} = \psi\frac{w_2^2}{2g} \qquad (1\text{-}8)$$

where: "a" = contraction loss coefficient defined as the ratio of the minimum flow cross section to the cross section of the smaller pipe

w_2 = average velocity in the smaller section.

Coefficients a and ψ depend on the ratio of the pipe's cross sections. Typical values are given in Table 1-3. Coefficient ψ can also be estimated from the following formula:

$$\psi = \frac{1.5(1 - A_2/A_1)}{3 - A_2/A_1} \qquad (1\text{-}9)$$

For a sudden expansion, an approximate expression from the Bernoulli equation is:

$$h_\ell = \frac{(w_1 - w_2)^2}{2g} \qquad (1\text{-}10)$$

Gradual expansion and contractions are illustrated in Figure 1-4. For smooth conical expansions (for $7° < \beta < 35°$, see Figure 1-4A):

$$h_\ell = 0.35\left(\log\frac{\beta}{2}\right)^{1.22}\frac{(w_1 - w_2)^2}{2g} \qquad (1\text{-}11)$$

Table 1-3
Coefficients of Contraction for Use in Equation 1-8

A_1/A_2	a	ψ
0.01	0.6	0.5
0.1	0.61	0.46
0.2	0.62	0.42
0.4	0.65	0.33
0.6	0.7	0.23
0.8	0.77	0.13
1.0	1.0	0.0

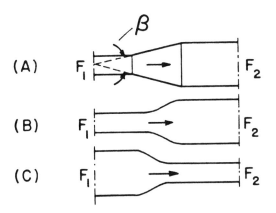

Figure 1-4. Flow through gradual expansions and contractions.

At $\beta > 40°$, head losses may be very high and even exceed those in sudden expansions.

For gradual contractions as in Figure 1-4C, head losses are very small. Equation 1-8 can be used allowing $h_\ell = 0.05$, independent of the ratio of A_2/A_1, provided that the flow is turbulent in the narrow cross section. If the flow in the contracted section is laminar, a pressure decrease generally is observed that does not follow Poiseuille's law. This decrease occurs over a length equivalent to $0.065 ReD$; the entrance region of the pipe. The pressure gradient at the entrance of a pipe of length L can be estimated from the data given in Table 1-5.

An alternate approach to estimating friction losses for gradual expansions with turbulent flow is to use the following equation:

$$h_\ell = C_L \frac{(w_1 - w_2)^2}{2g} \tag{1-12}$$

Table 1-4
Values for Estimating ΔP for Laminar
Flow in Gradual Contractions

Re L/D	0.005	0.01	0.02	0.03	0.04	0.05	0.06
$(P_0 - P_1)/(\gamma w^2/2g)$	2.1	2.6	3.4	4.1	4.7	5.3	6.0

D = pipe diameter; L = pipe length; γ = liquid specific gravity

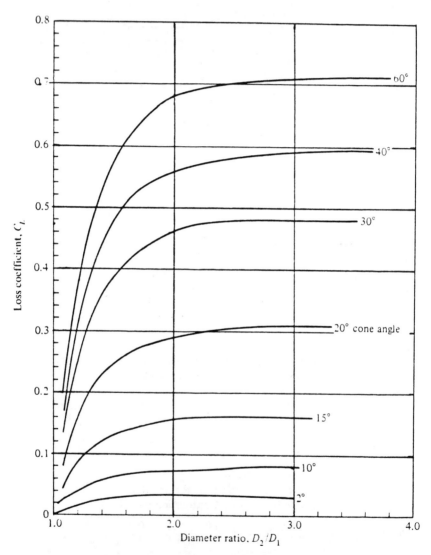

Figure 1-5. Loss coefficients for gradual expansions[2, 6].

Values of the loss coefficient C_L as a function of diameter ratios and different cone angles are given in Figure 1-5.

Pipe Network Analysis

This subsection outlines a unified basis for evaluating pipe network systems. Computations often require a trial and error solution and are best performed on a computer, or for simple systems, a hand calculator.

To evaluate flowrates and head losses through a pipe network, the system should be divided into a series of loops. A system of equations can be prepared to describe the conditions in each section.

The Hardy-Cross method is based on the following general formula:

$$\Delta Q_\ell^{(m+1)} \;=\; \Delta Q_\ell^{(m)} - \frac{F_\ell^{(m)}}{dF_\ell^{(m)}/d(\Delta Q_\ell)} \tag{1-13}$$

The following procedure is used:
1. Assume an initial flowrate for each pipe loop such that all junction continuity equations are satisified.
2. Compute the sum of the head losses around a loop of the network. Proper signs to signify flow direction must be maintained. For example, if the flow direction in one line is opposite to the flow in the loop, then head loss h_f is negative.
3. Accumulate absolute values of $n_i k_i Q_i^{n_i-1}$ around the same loop (explained below).
4. Evaluate ΔQ from the formula below.
5. Repeat steps 2 through 4 for each network loop.
6. Repeat steps 2 through 5 until computed ΔQ's converge.

The initially assumed flowrates in all pipes in the loop are adjusted when computing ΔQ values. Each equation ($F_\ell = 0$) is evaluated with all Q's summing to zero. That is, $\Delta Q^{(m)} = 0$, hence:

$$\Delta Q \;=\; \frac{-F_\ell}{dF_\ell/d(\Delta Q_\ell)} \tag{1-14}$$

The head loss around each loop is:

$$F_\ell \;=\; \Sigma k_i Q_i^{n_1} \tag{1-15}$$

The derivative of F_ℓ is:

$$\frac{dF_\ell}{d\Delta Q_\ell} = \Sigma \mid n_i K_i Q_i^{n_i-1} \mid \tag{1-16}$$

Combining the above expressions gives:

$$\Delta Q = -\frac{\Sigma K_i Q^{n_i}}{\Sigma \mid n_i K_i Q_i^{n_i-1} \mid} \tag{1-17}$$

The Hazen-Williams equation is used to define the loss factor K and the exponential term, n:

$$Q = -\frac{\Sigma(K_i)Q_i^{1.852}}{1.852\Sigma \mid (K_i)Q_i^{0.852} \mid}$$

$$= -\frac{\Sigma(h_{fi})}{1.852\Sigma \mid (h_f/Q)_i \mid} \tag{1-18}$$

An alternate converging scheme is based on the Newton-Raphson method. This scheme starts with a flow estimate and iteratively computes better estimates by means of quadratic convergence. The equation containing the unknown is expressed as a function that is set equal to zero when convergence is attained:

$$F(x) = 0 \tag{1-19}$$

Better estimates of unknown parameter x (head loss or volumetric flow Q) are computed by:

$$x^{(m+1)} = x^{(m)} - \frac{F(x^{(m)})}{df^{(m)}/dx} \tag{1-20}$$

Superscripts in Equation 1-20 denote number of iterations; not exponents.

The method may be used to solve any set of equations describing flow in pipe networks (i.e., flowrate in each pipe section, head at each junction, corrective flowrate around each loop).

In vector notation:

$$\vec{x}^{(m+1)} = \vec{x}^{(m)} - D_j^{-1}\vec{F}(x^m) \tag{1-21}$$

where $\vec{x}$ and $\vec{F}$ are unknown vectors; D_j^{-1} is the inverse of the Jacobian (i.e., D_j^{-1} replaces $1/dF/dx$).

In matrix form, a system of general equations is:

$$\vec{H} = \begin{vmatrix} H_1 \\ H_2 \\ \vdots \\ H_n \end{vmatrix} \quad \text{or} \quad \Delta\vec{Q} = \begin{vmatrix} \Delta Q_1 \\ \Delta Q_2 \\ \vdots \\ \Delta Q_n \end{vmatrix} \tag{1-22}$$

where: H = system of equations for head losses
        Q = system of equations for volumetric flowrate.

For a system of head losses, the Jacobian is:

$$D_j = \begin{pmatrix} \dfrac{\partial F_1}{\partial H_1} & \dfrac{\partial F_1}{\partial H_2} & \cdots & \dfrac{\partial F_1}{\partial H_n} \\[2ex] \dfrac{\partial F_2}{\partial H_1} & \dfrac{\partial F_2}{\partial H_2} & \cdots & \dfrac{\partial F_2}{\partial H_n} \\[2ex] \vdots & \vdots & & \vdots \\[2ex] \dfrac{\partial F_m}{\partial H_1} & \dfrac{\partial F_m}{\partial H_2} & \cdots & \dfrac{\partial F_m}{\partial H_n} \end{pmatrix} \tag{1-23}$$

Since division by a matrix is undefined (see Equation 1-21), a solution vector must be obtained, where:

$$D_j\vec{Z} = \vec{F} \tag{1-24}$$

Hence:
$$\vec{H}^{(m+1)} = H^{(m)} - \vec{Z}^{(m)} \tag{1-25}$$

The Newton-Raphson method obtains the solution to a system of non-linear equations via iteratively solving a system of linear equations. See Reference 4 for sample calculations using the Hardy-Cross method and further explanation of the Newton-Raphson method.

Channel Flow

The following formulas provide estimates of flow capacity over different types of weirs.

For flow over rectangular weirs having end contractions:

$$Q(cfs) = 3.33(L - 0.2H)H^{3/2}$$
$$Q(MGD) = 0.646317Q(cfs) \tag{1-26}$$

where: L = length of weir crest (ft)
H = head (ft)

For flow over Cipolletti (trapezoidal) weirs, the volumetric flowrate, given in cfs, can be computed from:

$$Q(cfs) = 3.367LH^{3/2} \tag{1-27}$$

For rectangular (suppressed) weirs without end contractions, the volumetric flowrate is:

$$Q(cfs) = 3.33LH^{3/2} \tag{1-28}$$

For water discharge rates over a 30° V-notch weir:

$$Q(cfs) = 0.670H^{5/2} \tag{1-29}$$

For discharge rates over a 45° V-notch weir:

$$Q(cfs) = 1.035H^{5/2} \tag{1-30}$$

For discharge rates over a 60° V-notch weir:

$$Q(cfs) = 1.443H^{5/2} \tag{1-31}$$

For discharge rate over a 90° V-notch weir:

$$Q(cfs) = 2.50H^{5/2} \tag{1-32}$$

For a discharge rate over a 120° V-notch weir:

$$Q(cfs) = 4.33H^{5/2} \tag{1-33}$$

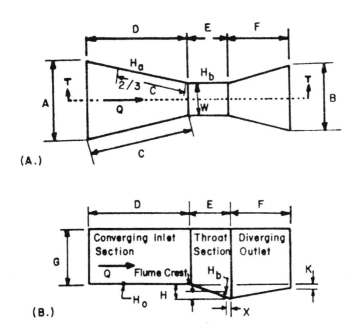

Figure 1-6. Details of a Parshall flume: (A) plan view (B) sectional view.

Table 1-5 provides size information on Parshall flumes. Definitions of major dimensions are given in Figure 1-6.

For Parshall flumes having throat widths ranging from 1 to 18 inches, use the following formulas:

$$
\begin{aligned}
1 \text{ in., } Q(\text{cfs}) &= 0.338H^{1.55} \\
2 \text{ in., } Q(\text{cfs}) &= 0.676H^{1.55} \\
3 \text{ in., } Q(\text{cfs}) &= 0.922H^{1.547} \\
6 \text{ in., } Q(\text{cfs}) &= 2.06H^{1.58} \\
9 \text{ in., } Q(\text{cfs}) &= 3.07H^{1.53}
\end{aligned}
\tag{1-34}
$$

For widths ranging from 1 to 8 ft:

$$
Q(\text{cfs}) = 4WH^{1.522}W^{0.26}
\tag{1-35}
$$

Table 1-5
Dimensions and Capacities for Parshall Flumes

Throat Width (W)	Dimensions in Feet and Inches												Free Flow Capacities			
													Minimum		Maximum	
ft/in.	A	B	C	2/3 C or 2/3(W/244)	D	E	F	G	H	K	X	Y	cfs	mgd	cfs	mgd
1"	6-19/32"	3-21/32"	1' 2-9/32"	9-17/32"	1' 2"	3"	8"	6"	1-1/8"	3/4"	5/16"	1/2"	0.01	0.006	0.2	0.13
2"	8-13/32"	5-5/16"	1' 4-5/16"	10-7/8"	1' 4"	4-1/2"	10"	8"	1-11/16"	7/8"	5/8"	1"	0.02	0.012	0.4	0.26
3"	10-3/16"	7"	1' 6-3/8"	1' 1-1/4"	1' 6"	6"	1'	1' 3"	2-1/4"	1"	1"	1-1/2"	0.03	0.02	0.6	0.39
6"	1' 3-1/2"	1' 3-1/2"	2' 7/16"	1' 4-5/16"	2'	1'	2'	1' 6"	4-1/2"	3"	2"	3"	0.05	0.03	2.9	1.9
9"	1' 10-5/8"	1' 3"	2' 10-5/8"	1' 11-1/8"	2' 10"	1'	1' 6"	2'	4-1/2"	3"	2"	3"	0.1	0.06	5.1	3.3
12"	2' 9-1/4"	2'	4' 6"	3'	4' 4-7/8"	2'	3'	3'	9"	3"	2"	3"	0.4	0.26	16.	10.
18"	3' 4-3/8"	2' 6"	4' 9"	3' 2"	4' 4-7/8"	2'	3'	3'	9"	3"	2"	3"	0.5	0.32	24.	15.
24"	3' 11-1/2"	3'	5'	3' 4"	4' 10-7/8"	2'	3'	3'	9"	3"	2"	3"	0.7	0.46	33.	21.
3'	5' 1-7/8"	4'	5' 6"	3' 8"	5' 4-3/4"	2'	3'	3'	9"	3"	2"	3"	1.0	0.65	50.	32.
4'	6' 4-1/4"	5'	6'	4'	5' 10-5/8"	2'	3'	3'	9"	3"	2"	3"	1.3	0.84	68.	44.
6'	8' 9"	7'	7'	4' 8"	6' 10-3/8"	2'	3'	3'	9"	3"	2"	3"	2.6	1.7	104.	67.
8'	11' 1-3/4"	9'	8'	5' 4"	7' 10-1/8"	2'	3'	3'	9"	3"	2"	3"	4.6	2.0	140.	90.
10'	15' 7-1/4"	12'	14' 3-1/4"	6'	14'	3'	6'	4'	1' 1-1/2"	6"	1'	9"	6.0	3.9	200.	129.
15'	25'	18' 4"	25' 6"	7' 8"	25'	4'	10'	6'	1' 6"	9"	1'	9"	8.0	5.2	600.	388.
20'	30'	24'	25' 6"	9' 4"	25'	6'	12'	7'	2' 3"	1'	1'	9"	10.0	6.5	1000.	646.
25'	35'	29' 4"	25' 6"	11'	25'	6'	13'	7'	2' 3"	1'	1'	9"	15.0	9.7	1200.	775.
30'	40' 4-3/4"	34' 8"	26' 6-1/4"	12' 8"	26'	6'	14'	7'	2' 3"	1'	1'	9"	15.0	9.7	1500.	969.
40'	50' 9-1/2"	45' 4"	27' 6-1/2"	16'	27'	6'	16'	7'	2' 3"	1'	1'	9"	20.0	13.0	2000.	1293.
50'	60' 9-1/2"	56' 8"	27' 6-1/2"	19' 4"	27'	6'	20'	7'	2' 3"	1'	1'	9"	25.0	16.0	3000.	1939

For widths greater than 10 ft:

$$Q(cfs) = (3.6875W + 2.5)H^{1.6} \qquad (1\text{-}36)$$

References

1. *Chem. Engr.*, 75 (13):198–199 (1968).
2. Crane Co., Technical Paper No. 410, "Flow Through Valves, Fittings and Pipe," (1970).
3. Cheremisinoff, N. P., *Pocket Handbook of Flow Calculations*, Gulf Publishing Co., Houston, TX (1984).
4. Cheremisinoff, N. P., *Fluid Flow: Pumps, Pipes and Channels*, Ann Arbor Science Pub., Ann Arbor, MI (1981).
5. Cheremisinoff, N. P. and D. Azbel, *Fluid Mechanics and Unit Operations*, Ann Arbor Science Pub., Ann Arbor, MI (1983).
6. Folsom, R. G., Trans. Am. Soc. Mech. Engrs., 78, 1447–1460 (1956).

2

LIQUID FILTRATION BASICS

Suspended Solids Removal

The primary applications discussed in this manual pertain to wastewater treatment. Total solids (TS) in wastewater exist in a distribution of sizes. Analytical methods are available to distinguish the suspended fraction of the total solids and to further distinguish the settleable fraction within the suspended solids (SS). See References 1-4.

Processes for SS removal fall into three categories: pretreatment to protect subsequent processes, treatment to reduce effluent concentrations to specified standards, and solids removal to produce concentrated recycle streams to maintain other processes.

Table 2-1 lists common SS removal methods along with typical loading ranges.

Only mechanical techniques are summarized in this manual for solids removal. All processes may be symbolically represented by the material balance system in Figure 2-1. Areas of rectangles denote volumes of substances at various separation stages. This figure illustrates the method of material balance record keeping for analyzing solids removal process schemes. Define AEFB as a quantity of material of mass M_H, volume V_H and density ρ_H.

The material is prepared in granular form and then mixed with a quantity of liquid ECDF of volume V_c, mass M_c and density ρ_c. If no chemical reactions take place, then the mass and volume of the mixture ACDB will be equal to the mass and volume of the initial components. Upon flowing through subprocess (A), the system divides into two parts: sludge AMNB and filtrate (if filtration) or decanted liquid (if sedimentation) MCDN. The sums of the volumes and masses of the separation products equal those of the initial suspension. The concen-

Table 2-1
Typical SS Separation Process Applications[4]

| Type of Separation Process | Application | Typical Loading Ranges | | | Expected Effluent SS |
		Hydraulic gpm/sq ft	Infl. Solids mg/l	lb/day/sq ft	mg/l
Straining					
Wedge Wire Screens	Preliminary Treatment of Raw Wastewater	10-30	200	25-75	150-190
Microscreens	Polishing of Biological System Effluent	3-10	30	1-2	5-15
Gravity Separation					
Plain Sedimentation	Primary Treatment	0.4-1.6	200	0.5-2	120-80
Chemical Coagulation and Settling	Chemical Treatment of Raw Sewage (Phosphate Removal Levels)	0.3-1.0	200(c)	1-6	20-60
Plain Sedimentation (secondary)	Separation of Solids after Activated Sludge Treatment	0.25-0.75	2000-5000	4-40	10-50
Granular Media Filtration	Polishing of Biological Effluent or Filtration of Chemically-Coagulated and Settled Raw Wastewater or	4-8	30	1-2	5-15
		3-5	40	1-2	10-20
	Secondary Effluent	3-5	5-10	1-2	1-3

trations of the sludge, η_0, is higher than that of the initial suspension, and the concentration of the filtrate, η_f, is less than that of the suspension, i.e.:

$$\eta_0 > \eta_f.$$

If all the solid particles contained in the sludge and filtrate are extracted and then compressed to the density of the solid lumps of the initial material, the thickened sludge would have volume of solid particles ARTB, with density ρ_H, mass $M_{H,0}$ and volume $V_{H,0}$, and the filtrate volume MQPN with $M_{H,\phi}$ $V_{H,\phi}$ and ρ_H. If the operation is 100% efficient, the masses and volumes of the solid particles in the sludge (ARTB and MQPN), as well as in the filtrate (RMNT and QCDP), will be equal correspondingly to those of AEFD and ECDF.

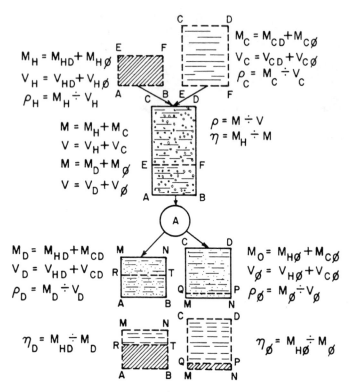

Figure 2-1. Generalized process material balance scheme.

There are nine values represented by two unit measures—mass and volume. This means a total of 18 values for which 12 material balance equations can be written. These equations are written next to the appropriate stage in the scheme. To solve this matrix of equations, six more expressions are required. Additional expressions are often those that characterize the physical properties of the substances (densities, specific volumes) and the compositions of the mixtures.

Examples of mass concentration, η, and density, ρ, expressions are given in Figure 2-1. In performing calculations, different variations of material balances usually are not necessary. Normally, only two or three equations are required to evaluate specific parameters of interest.

The exact expressions required depend on the information known about the system and the parameters to be evaluated.

For filtration processes, we may assume that the concentration of solids in the filtrate, η_ϕ, is zero; i.e., the entire solid phase of the suspension is transformed into a sludge. In this case, $M_H = H_{H,0}$ and $M_{H,d} = 0$.

One of the principle mechanical operations for solids removal is filtration. The balance of this chapter provides general notes on filtration theory and practice.

Flow Through Fixed Beds

Flow through packing arrangements has been extensively studied by Carman[5]. The friction factor at low Re is:

$$f = 90(1 - \epsilon)^2/\epsilon^3 Re \qquad (2\text{-}1)$$

The correlation shows a $\pm 10\%$ scatter of data.

The frictional pressure drop ΔP in a stream of fluid flowing through length L of packed bed of uniform spheres is described by the general expression:

$$\frac{\Delta P}{L} = \frac{2f\rho_f u^2}{gD_p} \qquad (2\text{-}2)$$

where: u = superficial velocity
D_p = particle diameter

Where the sphere diameter D_p is more than 10% of the bed width, Equation 2-2 overpredicts ΔP, and appropriate wall-effect correction factors must be applied. For beds of randomly spaced spheres, voidage ϵ ranges from 0.38 to 0.47. The lower voidage is that observed by a random densely packed bed of uniform spheres.

From dimensional analysis, the pressure drop in a fluid traversing geometrically similar beds can be expressed by:

$$\frac{\Delta P}{L} \simeq \mu_f^{2-n} u^n d^{n-3} \rho_f^{n-1} \qquad (2\text{-}3)$$

where d is a representative bed dimension chosen as a pore length analogous to the hydraulic radius of a conduit.

$$d = \frac{\text{Mean cross-sectional area of flow channels through bed}}{\text{Mean wetted perimeter of flow channels}}$$

$$= \frac{(\text{total bed volume})}{(\text{total bed surface})} = \epsilon/s \qquad (2\text{-}4)$$

where: s = surface of solids per unit bed volume

For uniform spheres:

$$s = 6(1 - \epsilon)/D_p$$
$$d = \epsilon D_p/6(1 - \epsilon) \qquad (2\text{-}5)$$

The ΔP and f expressions can be applied to other regular geometries (e.g., rings, saddles, wire crimps) whose surfaces can be determined accurately, provided D_p is replaced by $6V_p/s_p$; where V_p = particle volume, s_p = particle surface. These equations can also be applied to mixtures of particles of various sizes and shapes.

For ϵ in the range of 0.26 to 0.89:

$$\frac{\Delta P}{L} = \frac{180(1 - \epsilon)^3 \mu_f u}{g\epsilon^3 D_p^2}$$

$$= \frac{5(1 - \epsilon)^2 \mu_f u}{g\epsilon^3 (V_p/s_p)^2} \qquad (2\text{-}6)$$

Equation 2-6 is known as the Carman-Kozeny equation [5,6]. It is derived by assuming that a granular bed is equivalent to a group of parallel similar channels, such that the total internal surface and volume are equal to the bed's particle surface and bed's voidage volume, respectively.

The general expression for streamline flow through a uniform channel is:

$$\frac{\Delta P}{L_c} = \frac{k\mu_f u}{g R_H^2} \qquad (2\text{-}7)$$

where: k = coefficient
               = 2 for circle
               = 1.78 for square
               = 2.65 for rectangle with length equal to 10 times the side
               = 3 for rectangle of infinite length
     L_c = channel length

Equation 2-7 can also be stated as:

$$\frac{\Delta P}{L} = \frac{k(L_c/L)^2(1 - \epsilon)^2\mu_f u}{g\epsilon^3(V_p/s_p)^2} \tag{2-8}$$

For actual beds, $k(L_c/L)^2$ has a value of about 5 and $k \simeq 2.5$.

From pressure-drop measurements through beds of irregular particles, the specific surface s_p/V_p can be estimated:

$$s_p/V_p = \sqrt{\frac{g\Delta P\epsilon^3}{5(1 - \epsilon)^2\mu_f uL}} \tag{2-9}$$

Darcy's law relates the volumetric flowrate Q of a fluid flowing linearly through a porous medium directly to the energy loss, inversely to the length of the medium and proportionally to a factor called the hydraulic conductivity K'.

$$Q = \frac{K'A(h_1 - h_2)}{\Delta h} \tag{2-10}$$

where: $\Delta h = \Delta z + \frac{\Delta P}{\rho} + \text{constant}$ (2-11)

Hydraulic conductivity K' depends on the properties of the fluid and the pore structure of the medium. The hydraulic conductivity is temperature-dependent. In terms of the intrinsic permeability k' and fluid properties:

$$K' = \frac{k'\rho g}{\mu} \tag{2-12}$$

k' is a function only of the pore structure and is not temperature dependent. In differential form, Darcy's equation is:

$$\frac{Q}{A} = q = -\frac{k}{\mu}\frac{dP}{dx}$$ (2-13)

where: q = seepage velocity (equivalent to the velocity of approach V_∞)

The Reynolds number is defined in this case as:

$$Re_p = \frac{D_p V_\infty \rho}{\mu(1 - \epsilon)}$$ (2-14)

Filtration Process

Liquid filtration is a major operation in the treatment of water for achieving the supplemental removals of suspended solids from wastewater effluents of biological and chemical treatment processes and in product or feedstock recovery. Filtration is also used to remove chemically precipitated phosphorus and iron.

In waste treatment a major design issue is whether filtration can meet required specified effluent quality goals. If the goal is to upgrade the effluent of an existing secondary treatment works, the present performance and the reasons for that performance must be evaluated. For example, what portions of the present effluent BOD (Biological Oxygen Demand) are of soluble and suspended origin? The filter can remove only a portion of the suspended BOD. If the effluent contains highly soluble BOD, the solution may be limited to upgrading the secondary treatment. If the effluent contains primarily suspended BOD, effluent filtration or upgrading the secondary settling will be possible alternative solutions. The performance of the granular filters can be estimated from performance at similar plants elsewhere, or by pilot studies at the plant.

After effluent quality targets are established, the following must be considered:

1. What are the appropriate flow schemes?
2. What minimum filter run length is acceptable?
3. What filter configurations are appropriate for wastewater?
4. Is pilot scale testing needed, and if so, how should it be conducted?

5. What filtration rate and terminal headloss should be provided?
6. What filter media size and depth should be provided?
7. Should gravity or pressure filters be provided?
8. What system of flow control should be used?
9. What backwash provisions are needed for each filter media alternative being considered to ensure effective backwashing?
10. What underdrain system is appropriate for the media and backwash regime intended?

The two main operations performed are *filtration* and *backwashing*. Filtration is accomplished by passing the fluid to be filtered through either a filter bed composed of granular material or a porous media (e.g., cloth) with or without the addition of chemicals. Within a granular filter bed, the removal of the suspended solids contained in the wastewater is accomplished by a complex process involving one or more removal mechanisms, such as straining, interception, impaction, sedimentation, and adsorption. The end of the filter run (filtration phase) is reached when the suspended solids in the effluent start to increase (breakthrough) beyond an acceptable level, or when a limiting head loss occurs across the filter bed (see Figure 2-2). Both these events generally occur simultaneously.

If either of these conditions are reached, the filtration phase is terminated, and the filter must be backwashed to remove suspended solids that have accumulated within the filter media. This is done by reversing the flow through the filter. In a granular bed this is accomplished by providing a sufficient flow of wash water to fluidize the granular filtering medium. The wash water moving past the medium will shear away the material attached to the individual grains of the granular medium. During backwashing, care should be taken not to expand the bed to such an extent that the effectiveness of the shearing action of the wash water is reduced. In most wastewater-treatment plants, the wash water containing the suspended solids that are removed from the filter are returned to the primary settling facilities.

There are generally two primary means by which particles are removed from the stream: *entrapment* and *adhesion*.

Entrapment includes such mechanisms as straining, interception, inertial impact, sedimentation, and other hydrodynamic forces. These "mechanical" retention methods are related to the effective size, the uniformity coefficient, and the depth of the particular media. These factors determine the void areas or flow channels between the grains of media when the filter bed is in place.

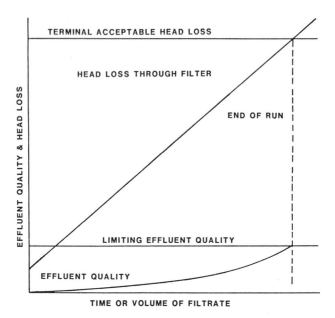

Figure 2-2. Illustrates run cycle for a filtration process.

The size of these channels determines in a large part the size of particles which can be removed by mechanical means. If the particles are granular in nature, there is a certain size above which 100% filtration efficiency will be achieved. However, most applications deal with particles (oil, silt, floc, etc.) which will deform under the hydraulic forces to which they are subjected as they pass through the media bed. Consequently, straining alone is not an efficient means of filtration.

Some particles are not susceptible to straining and are then removed by interception at a point of impingement on the media grains. Others are retained by inertial impact upon the surface of the media. In any media configuration, there are also areas that are "sheltered" from the flow path. As particles enter these areas they are subject to the effects of sedimentation and become removed from the stream. The effectiveness of these mechanisms is determined by the combination of the nature and amounts of contaminants to be removed and the size and type of media used.

Adhesion is another mechanism that affects filtration efficiency. Adhesion may consist of combinations of chemical bonding, Van der Waals forces, electrostatic forces, or mutual adsorption.

To remove particulates entirely by mechanical means a very fine media must be used and filtration cycles should be short. If the media is porous enough to substain good run lengths, the filter efficiency will suffer as a result. As an example, 0.3 mm sand typically used for single media filtration is approximately 300 microns in diameter. The smallest void area between any three grains is approximately 40–60 microns. There will be smaller openings where two grains come together, but most of the void areas will be in this 40–60 micron range. However, the effluents that are encountered will have a particle distribution from submicron to approximately 150 micron. A typical particle distribution for waste waters can often be bimodal. Some of these particles will be removed mechanically, but most would find their way through the media bed if mechanical separation mechanisms were the only ones employed.

In water, most solids have a naturally occurring electrical charge. This charge is almost always negative. It is the physio-chemical and molecular forces that are created by these charged particles that cause adhesion to take place.

These forces increase by the square of the distance of approach, so the nearer the particle comes to the media grain, the greater the adhesion force which is exerted on the particle. Here the mechanical elements of removal become significant. Inertial impact, interception, and sedimentation all work to bring the particle into close proximity with the media grain so the adhesive forces can hold that particle. Other physical properties also come into play: the viscosity of the liquid (higher viscosity has the effect of making it harder for the particle to move out of the flow path of the liquid), and centrifugal force (as the flow path of the liquid moves around the spherical media grains there is a centrifugal force added to the inertia of the particle). Both forces tend to move the particle out of the flow path of the liquid as it passes through the filter media.

Although most particles in water have a negative charge, the media also often has a negative charge. These like charges tend to repel each other and if they are strong enough, little or no adhesion takes place. This is especially true of colloidal suspensions. Although colloidal particles tend to move across the shear planes within the liquid and thereby come in close proximity of the media grains, these particles can only be removed in filters where the electro-kinetic charges of the particles and the media are of opposite signs.

To promote filtration efficiency through adhesion, it is often necessary to provide for the addition of polyelectrolytes to the stream. These polymers can be used to provide a variety of effects. These effects include floc formation, molecular bridging, charge reversal, etc. Of these, floc formations are probably the most commonly used but are least desirable because they involve large dosages of polymer, which tends to form flocs that blind the filter media in a relatively short time. It also produces large amounts of sludges which are hard to dewater. Molecular bridging involves the addition of polymers whose molecules attach onto two or more colloidal particles forming an aggregate of the particles and the polymer. The aggregate is distinguished from the floc by its smaller size, increased physical stability and relative ease of dewatering. Although the particles and the polymer are all negatively charged, the chemical bonds overcome the electro-kinetic forces keeping the particles apart. Charge neutralization is the most effective means of chemical treatment. Depending on the application, the polymer may be added to change the charge of either the particles or the media. In most cases, however, the short contact time and wide dispension of polymer lends itself to the changing of the media charge from negative to positive. This allows the strong electro-kinetic adhesion forces to remove the negatively charged particles.

The two major types of filtration are "cake" and "filter-medium" filtration. In the former, solid particulates generate a cake on the surface of filter medium. In filter-medium filtration (sometimes called blocking or clarification), solid particulates become entrapped within the complex pore structure of the filter medium. The filter medium for the latter case consists of cartridges or granular media. Examples of common media are sand, diatomite, coal, cotton or wool fabrics, metallic wire cloth, porous plates of quartz, chamotte, sintered glass, metal powder, and powdered ebonite. The average pore size and configuration (including tortuosity and connectivity) are established from the size and form of individual elements from which the medium is manufactured. On the average, pore sizes are greater for larger medium elements. In addition, pore configuration tends to be more uniform with more uniform medium elements. The fabrication method of the filter medium also affects pore size and form. See Cheremisinoff et al.[7] for a discussion of filtration mechanisms.

References

1. *Standard Methods for Examination of Water & Wastewater,* 13th Edition, American Public Health Assoc., NY (1971).
2. Rickert, D. A., and J. V. Hunter, "General Nature of Soluble and Particulate Organics in Sewage and Secondary Effluent," *Water Research,* 5, 421 (1971).

3. Hunter, J. V. and H. Heukelekian, "The Composition of Domestic Sewage Fractions," *JWPCF*, 37, 1142 (Aug. 1965).
4. *Process Design Manual for Suspended Solids Removal,* U.S. Environmental Protection Agency, EPA 625/1-75-003a (Jan. 1975).
5. Carman, P. C., Trans. Inst. Chem. Engrs., (London), 15, 150 (1937); *J. Soc. Chem. Ind.* (London), 57, 225 (1938).
6. Kozeny, J., Sitzber, Akad. Wiss. Wien, *Math-natur.* k1 (Abt. 11-a, 136, 271 (1927).
7. Cheremisinoff, N. P. and D. Azbel, *Liquid Filtration*, Ann Arbor Science Pub., Ann Arbor, MI (1983).

3

CAKE FILTRATION LAWS

Filtration Rate

The theory of cake filtration is discussed in detail by Cheremisinoff et al.[1], and further references are given in that volume. Only a summary of principle design equations are given here.

Since pore sizes in a cake and filter medium are small, the flow regime is laminar and Poiseuille's law applies. Filtration rate is directly proportional to the pressure drop and inversely proportional to fluid viscosity and the hydraulic resistance of the cake and filter medium.

$$u = \frac{1}{A} \frac{dV}{d\tau} = \frac{\Delta P}{\mu(R_c + R_f)} \tag{3-1}$$

where: V = filtrate volume
A = filter area
τ = filtration time
μ = filtrate viscosity
ΔP = pressure difference
R_c = filter cake resistance
R_f = initial filter resistance (resistance of filter plate and filter channels)
u = filtration rate; i.e., filtrate flow through cake and plate

R_f is often assumed to be a constant. Filter cake resistance R_c is the resistance to filtrate flow per unit area of filtration; R_c increases with increasing cake thickness.

The ratio of cake volume to filtrate volume is x_o. Cake volume is $x_o V$. For a cake height, h_c:

$$x_o V = h_c A \qquad (3\text{-}2)$$

Cake resistance is:

$$R_c = r_o x_o \frac{V}{A} \qquad (3\text{-}3)$$

where r_o = specific volumetric cake resistances (m^{-2}).
Equation 3-1 can be stated in terms of the above definitions:

$$\frac{1}{A}\frac{dv}{d\tau} = u = \frac{\Delta P}{\mu(r_o x_o (V/A) + R_f)} \qquad (3\text{-}4)$$

x_o can also be expressed in terms of the mass ratio of solid particles settled on the filter plate to the filtrate volume (x_w). Also, instead of r_o, a specific mass cake resistance r_w can be used, where r_w is the flow resistance resulting from a uniformly distributed cake in the amount of 1 kg/m^2.
For a negligible plate resistance $(R_f = 0)$:

$$r_o = \frac{\Delta P}{\mu h_c u} \qquad (3\text{-}5)$$

For highly compressible cakes, $r_o \geq 10^{12} m^{-2}$. The limiting case is no flow $(v = 0)$, which denotes the start of filtration, whence:

$$R_f = \frac{\Delta P}{\mu u} \qquad (3\text{-}6)$$

Constant Pressure Filtration

For constant pressure drop and temperature, the relationship between filtration time and filtrate volume is:

$$V^2 + 2\left(\frac{R_f A}{r_o x_o}\right)V = \left(2\,\frac{\Delta P A^2}{\mu r_o x_o}\right)\tau \qquad (3\text{-}7)$$

Filtration constants are defined as:

$$K_c = \frac{2\Delta PA^2}{\mu r_o x_o} \tag{3-8}$$

$$C = \frac{R_f A}{r_o x_o} \tag{3-9}$$

Therefore, Equation 3-7 is:

$$V^2 + 2VC = K_c \tau \tag{3-10}$$

K_c and C must be experimentally evaluated for the filter media and filtration conditions.

Equation 3-10 can also be stated as:

$$(V + C)^2 = K_c(\tau + \tau_o) \tag{3-11}$$

where:

$$\tau_o = \frac{C^2}{K_c} \tag{3-12}$$

Equation 3-11 is a parabolic relationship. A plot of $V + C$ vs. $\tau + \tau_o$ describes the filtration process (i.e., the characteristic filtration curve for the system).

Sample Calculation 3-1. A slurry is filtered in a bench scale filter press with a pressure drop of 0.3 psi and a temperature of 20°C. After 10 minutes, 4.9 liters of filtrate are obtained; after 20 minutes, 7.1 liters. Based on tests at other pressures, the cake compression coefficient is found to be s = 0.4. Determine the volume of filtrate after 40 minutes from a filter press having an area 10 times greater than the laboratory press and being operated at 2.0 atmospheres pressure. The liquid temperature is 55°C.

Solution:

First evaluate K_c and C:

$$v^2 + 2VC = K_c\tau$$
$$(4.9)^2 + 2(5.1)C = 10\ K_c$$
$$(7.1)^2 + 2(7.1)C = 20\ K_c$$
$$C = 0.39,\ K_c = 2.79$$

The filtration coefficients change as a function of process conditions (temperature, pressure). Resistance coefficient can be approximated from:

$$r_0 = a'\Delta P^s \qquad (3\text{-}13)$$

where: a' = coefficient; coefficient C can be expressed as:

$$C = \frac{R_f}{a'x_o}\frac{A}{\Delta P^s} \qquad (3\text{-}14)$$

$r/a'x_o$ is constant and independent of conditions. By ratio, we determine the constant at the higher pressure of 2.0 atm:

$$\frac{C_2}{C_1} = \frac{A_2}{A_1}\left(\frac{\Delta P_1}{\Delta P_2}\right)^s$$
$$A_2/A_1 = 10$$
$$s = 0.4$$

And solving for C_2:

$$C_2 = 0.39 \times 10\left(\frac{0.3}{2.0}\right)^{0.4} = 1.83$$

K_c is determined from:

$$K_c = \frac{2}{a'x_o}\frac{\Delta P^{1-s}A^2}{\mu}$$

$$\frac{K_{c2}}{K_{c1}} = \left(\frac{\Delta P_2}{\Delta P_1}\right)^{1-s} \left(\frac{A_2}{A_1}\right)^2 \frac{\mu_1}{\mu_2}$$

Viscosity of water:

$\mu_1 = 1$ cp @ 20°C
$\mu_2 = 0.5$ cp @ 55°C.

Solving for K_{c2}:

$$K_{c2} = 2.79 \left(\frac{2.0}{0.3}\right)^{1-0.4} \times 10^2 \times \frac{1}{0.5} = 1742$$

From Equation 3-12: evaluate τ_o:

$$\tau_o = \frac{(1.83)^2}{1742} = 1.92 \times 10^{-3}$$

Applying Equation 3-11:

$$(V + 1.83)^2 = 1742 (\tau + 1.92 \times 10^{-3})$$

where V is in liters and τ in minutes.

For $\tau = 40$ minutes:

$$V = [1742(40 + 1.92 \times 10^{-3})]^{1/2} - 1.83 = 2.62 \text{ liters.}$$

The final filtration rate can be approximated by:

$$\frac{dV}{d\tau} = \frac{K_c}{2(V + c)} \tag{3-15}$$

$$\frac{dV}{d\tau} = \frac{1742}{2(262 + 1.83)} = 3.30 \text{ liters/min}$$

The average filtration rate is:

$$\frac{V}{\tau} = \frac{262}{40} = 6.55 \text{ liters/min}$$

Constant Rate Filtration

During a constant rate process, pressure increases with cake thickness. The governing equation is:

$$\Delta P = \mu r_o x_o \left(\frac{V^2}{A^2 \tau}\right) + \mu R_f \left(\frac{V}{A\tau}\right) \tag{3-16}$$

The relationship between time and pressure for a constant rate filtration is:

$$\Delta P = \mu a x_o \Delta P^s \left(\frac{V}{A\tau}\right)^2 \tau + R_f \mu \left(\frac{V}{A\tau}\right) \tag{3-17}$$

At the start of filtration:

$$\Delta P_o = R_f \mu \left(\frac{V}{A\tau}\right) \tag{3-18}$$

And for an incompressible cake $(s = 0)$:

$$\mu a x_o \left(\frac{V}{A\tau}\right)^2 \tau + R_f \mu \left(\frac{V}{A\tau}\right) = \Delta P \tag{3-19}$$

For thick cakes, the filter medium resistance can be neglected $(R_f = 0)$

$$\Delta P^{1-s} = \mu x_o a \left(\frac{V}{A\tau}\right)^2 \tau \tag{3-20}$$

Additional useful formulas are
The weight of dry solids in the cake:

$$W = x_o V \tag{3-21}$$

where: x_o = weight of solids in cake per unit filtrate volume
The specific weight of the feed sludge:

$$\gamma = x_o (1 - mc)/c \tag{3-22}$$

where: c = concentration of solids in feed sludge
$\quad\quad\quad$ m = weight ratio of wet to dry cake

For a dilute suspension, c is small and $x_o \simeq c$.

Variable Rate and Pressure Filtration

The compressed force exerted on the cake section of a filter and corresponding to the local specific cake resistance $(r_w)_x$ is:

$$P = P_1 = P_{st} \tag{3-23}$$

where: P_1 = pressure exerted on sludge over entire cake thickness
$\quad\quad\quad$ P_{st} = static pressure over the same section of cake

At the sludge-cake interface:

$$P_{st} = P_1$$

$$P = 0$$

At the interface between the cake and filter plate:

$$P = P_1 - P_{st}'$$

where P_{st}' corresponds to the filter plate resistance, expressed as:

$$\Delta P_f = \mu R_f W \tag{3-24}$$

where: W = filtration rate $(m^3/m^2\text{-sec})$

The incremental dP across a cake of unit area is:

$$\frac{dq}{x_w dP} = \mu (r_w)_x W \tag{3-25}$$

Note that x_w is not sensitive to pressure variations. W is a constant for any cross section of cake. Equation 3-25 can be integrated over the cake thickness from 0 to q flow:

$$q = \frac{1}{\mu x_w W} \int_o^{P_1 - P'_{st}} \frac{dP}{(r_w)_x} \tag{3-26}$$

q and W are variables only when filtration conditions change. Coefficient $(r_w)_x$ is pressure dependent; the exact relationship must be evaluated from data obtained in a compression-permeability cell. Once this relationship is defined from a regression analysis of data, the integral in Equation 3-26 can be evaluated.

The relationship between W and P_1 can be established from pump characteristics. Filtration time can then be evaluated from:

$$\frac{dq}{d\tau} = W \tag{3-27}$$

or

$$\tau = \int_o^q \frac{dq}{W} \tag{3-28}$$

See Cheremisinoff et al.[1] for sample calculations.

Graphical Methods

The design equations outlined can be solved by graphical means to provide sizing criteria. This subsection briefly outlines the general approach. For details, see Cheremisinoff et al.[1]

The relationship between cake mass, cake thickness and filtrate volume can be derived from a material balance between the sludge solids being filtered and the solids comprising the cake.

$$W/S = W/S_c + \rho \nu \tag{3-29}$$

where: S, S_c = mass fractions of solids in the sludge and cake, respectively
ρ = density of the liquid
ν = filtrate volume per unit filter medium area
W = mass of dry solids per unit area of filter media.

The mass of dry cake is:

$$W = \frac{\rho S}{1 - S/S_c} \nu = c\nu \qquad (3\text{-}30)$$

where: c = fictitious concentration defined as the mass of dry solids per volume of filtrate

When $S << S_c$ (i.e., dilute sludges)

$$W = \rho S \nu \qquad (3\text{-}31)$$

The mass of particles deposited on the filter plate is related to the average porosity ϵ_{avg} and to the cake thickness L:

$$W = \rho_s(1 - \epsilon_{avg})L \qquad (3\text{-}32)$$

The relationship between average cake porosity and mass fraction of solids in the cake is:

$$S_c = \frac{\rho_s(1 - \epsilon_{avg})}{\rho_s(1 - \epsilon_{avg}) + \rho\epsilon_{avg}} \qquad (3\text{-}33)$$

For an indirect calculation of cake thickness L, where the filtrate volume is known:

$$L = \frac{S(\sigma(1 - S_c) + S_c)}{\sigma(S_c - S)} \nu \qquad (3\text{-}34)$$

or

$$L = \frac{\rho_s S_c}{\sigma(1 - S_c) + S_c} W \qquad (3\text{-}35)$$

where: $\sigma = \rho_s/\rho$

This last expression may be used to establish proper filter surface spacing.

At constant pressure drop across the cake, S_c and ϵ_{avg} are usually constant, and L has a linear dependence. When pressure drop ΔP_c varies over the cake, the porosity varies and this linear dependence no longer exists.

During filtration, the flow of filtrate through the porous filter cake and filter medium is laminar. Therefore, Darcy's law applies.

$$\frac{dP_L}{dx} = \frac{\mu q}{K} \tag{3-36}$$

where: K = permeability (Darcy's constant)
dP_L/dx = hydraulic gradient
μ = liquid viscosity
q = superficial liquid velocity

The actual average velocity in the pores of the cake and filter medium is q/ϵ.

As an approximation to Darcy's equation, the following modified version may be used:

$$\frac{dP_L}{dW} = \mu \alpha q \tag{3-37}$$

where: W = mass of dry solids deposited per unit area of filter medium
α = specific flow resistance

Replacing dP_L by $-dP_s$ (dP_s = cumulative drag pressure on solids, $P - P_L$; P = applied pressure; P_L = hydraulic pressure) and rearranging terms and integrating:

$$\mu q W = \int_0^{P-P_1} dP_s/\alpha \tag{3-38}$$

where P_1 = pressure required to overcome the resistance R_m of the medium (i.e., the filter plate). To evaluate the integral the resistances developed during cake filtration must be known. These are generally nonanalytical functions. The specific filtration resistance α can be related to the cumulative drag pressure on the solids P_s by an empirical relationship. Figure 3-1 is a plot of the specific filtration resistance vs. the

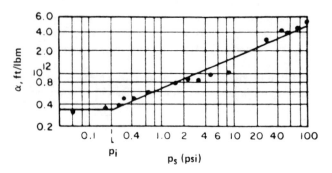

Figure 3-1. Plot of α vs. P_s.

cumulative drag pressure P_s, where α is a constant below some pressure P_i. The power functions of P_s are:

if $P_s < P_i$, then $\alpha = \alpha_i = aP_i^n$

if $P_s > P_i$, then $\alpha = aP_s^n$

The integral in Equation 3-38 is:

$$I = \int_o^{\Delta P_c} dP_s/\alpha \qquad (3\text{-}39)$$

where ΔP_c = pressure drop across cake ($P = P - P_i$).

$$\int_o^{\Delta P_c} \frac{dP_s}{\alpha} = \int_o^{P_i} \frac{dP_s}{\alpha_i} + \int_{Pi}^{\Delta P_c} \frac{dP_s}{aP_s^n} = \frac{P_i^{1-n}}{a} + \frac{\Delta P_c^{1-n} - P_i^{1-n}}{a(1-n)}$$

$$= \frac{\Delta P_c^{1-n} - nP_i^{1-n}}{a(1-n)} \qquad (3\text{-}40)$$

If n is < 0.6, the term P_i can be neglected for total pressures above 10 psi. For large n (> 0.7) the P_i term must be included. When n becomes large, the power function approximation is less accurate, and numerical methods should be employed for the integration.

When P_i is neglected:

$$I = \frac{\Delta P_c^{1-n}}{a(1-n)} \tag{3-41}$$

This gives:

$$c\mu q\nu = c\mu\nu\frac{d\nu}{dt} = \frac{(P-P_1)^{1-n}}{a(1-n)} \tag{3-42}$$

Filtration processes are classified according to the variation of the pressure and flowrate with time as established by the pumping mechanism. This serves as a basis for the following division of groups:

1. Constant-pressure filtration: the actuating mechanism is compressed air maintained at a constant pressure.
2. Constant-rate filtration: positive displacement pumps are employed in this group.
3. Variable-pressure, variable-rate filtration: the use of a centrifugal pump results in the rate varying with the backpressure on the pump.

The flowrate vs. pressure characteristics of these three types of filtering are summarized in Figure 3-2.

Since $P - P_1$ is approximately constant for most practical operations, Equation 3-43 is approximated by:

$$\int_0^\nu \nu d\nu = \frac{(P-P_1)^{1-n}}{ac\mu(1-n)} \int_0^t dt \tag{3-43}$$

Integrating this expression and noting that pressure P_1 required to overcome the resistance of filter plate R_m is $P_1 = \mu q R_m$, where $q = d\nu/dt$, the average specific filtration resistance is:

$$\frac{1}{\alpha_{avg}} = \frac{1}{P-P_1} \int_o^{P-P_1} dPs/\alpha \tag{3-44}$$

The derivation (see Cheremisinoff et al.[1] for details) results in:

$$q = \frac{d\nu}{dt} = \frac{P}{\mu(\alpha_{avg}W + R_m)} \tag{3-45}$$

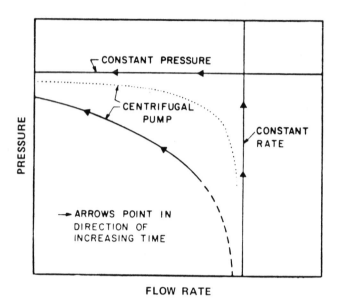

Figure 3-2. Effect of pump characteristics on filtration.

If the filter plate resistance R_m is small or the filtration lasts long enough for $\alpha_{avg}W$ to become much larger than R_m, P_1 can be neglected:

$$\frac{\nu^2}{2} = \frac{P^{1-n}}{ac\mu(1-n)}t \tag{3-46}$$

Equation 3-45 may be stated as (after integration):

$$\mu\alpha_{avg}c\frac{\nu^2}{2} + \mu R_m \nu = Pt \tag{3-47}$$

Combining Equations 3-46 and 3-47:

$$\alpha_{avg} = a(1-n)P^n \tag{3-48}$$

The constants α_{avg} and R_m can be evaluated from filtration data obtained at constant pressure. Rearranging:

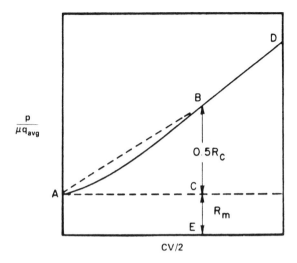

Figure 3-3. Graphical analysis for obtaining α_{avg} and R_m.

$$\frac{Pt}{\mu\nu} = \frac{P}{\mu q_{avg}} = \alpha_{avg}\frac{c\nu}{2} + R_m \qquad (3\text{-}49)$$

where the average flowrate over the entire filtration cycle is $q_{avg} = \nu/t$. A plot of $P/\mu q_{avg}$ vs $c\nu/2 = W/2$ is shown in Figure 3-3. The value of α_{avg} is given by:

$$\alpha_{avg} = \frac{(P/\mu q_{avg}) - R_m}{c\nu/2} \qquad (3\text{-}50)$$

and is denoted by BC/AC in Figure 3-3. In analyzing a constant pressure filtration operation it may be assumed that α_{avg} is constant. This is represented by line ABD in Figure 3-3.

The linearity of ABD depends on the relative pressure drops across the cake and filter plate. When ΔP_c rapidly approaches the applied pressure P_1, AB will represent only a small fraction of the total curve and can be ignored. Note that α_{avg} increases with time and approaches a constant value as ν increases indefinitely. The pressure drops across cake and filter plate are given by $\Delta P_c = \mu q \alpha_{avg}$ and $P_1 = \Delta P_m = \mu q R_m$, respectively. Equation 3-50 shows that $P/\mu q_{avg}$ is the sum of R_m

(line CE) and $R_c/2$ (line BC). The line BC is proportional to $\Delta P_c/2$ and EC is proportional to ΔP_m. The distance BC must be much larger than CE before the pressure drop across the cake approaches the applied pressure P. As long as ΔP_c changes, α_{avg} will also change.

Reference

1. Cheremisinoff, N. P. and D. S. Azbel, *Liquid Filtration,* Ann Arbor Science Publishers, Ann Arbor, MI (1983).

4

FILTRATION EQUIPMENT

Physical Straining Operations

Physical straining processes are those which remove solids by virtue of physical restrictions on a media which has no appreciable thickness in the direction of liquid flow. These operations may be grouped according to the nature of their straining action (See Table 4-1).

Among the most common systems are wedge wire screens (often inclined in orientation). These were originally developed for the pulp and paper industry to dewater and classify pulp slurries having solids contents of six percent or less. Units operate by gravity and function as an inclined drainage board with a screen of wedge wire construction having openings running transverse to the flow.

Screens usually consist of three sections with successively flatter slopes on the lower sections (See Figure 4-1). Screen wires are triangular in cross section. Some designs are straight and transverse to the flow.

Above the screen and running across its width is a headbox. A lightweight hinged baffle at the top portion of the screen reduces flow turbulence. To collect the solids coming off the end of the screen several arrangements can be used, including a trough with a screw conveyor.

Inclined screening units are generally constructed entirely of stainless steel. Lighter units with a fiberglass housing and frame cost about 25% less. Influent wastewater enters and overflows the headbox, on to the upper portion of the screen. On the screen's upper slope most of the fluid is removed from the influent. The solids concentrate on the successive slope, because it is flatter, and additional drainage occurs. On the screen's final slope the solids stop momentarily, simple drainage occurs, and the solids are displaced from the screen by oncoming solids.

Table 4-1
Physical Straining Processes for Wastewater Treatment[1]

Device	Range of Hydraulic Capacities	Straining Surface	Typical Waste Solids Composition	Typical Percent SS Removals
Inclined wedge-wire stainless steel screens	High flow rates 4-16 gpm/in of screen width	Coarse .01 to .06 in (250-1500 microns)	10 to 15% solids by weight	5-25
Rotary stainless steel wedge wire screens	16-112 gpm/sq ft	Coarse .01 to .06 in (250-1500 microns)	16 to 25% solids by weight	5-25
Centrifugal screens	40-100 gpm/sq ft	Medium 105	0.05-0.1%	60-70
Micro-Screens	Medium flow rates 3 to 10 gpm/sq ft	Medium 15-60 microns	250-700 mg/l (app. 0.05%)	50 to 80
Diatomite filters	Medium flow rates 0.5-1.0 gpm/sq ft	N/A		up to 90
Ultra-Filters	Low flow rates 5 to 50 gpd/sq ft	Fine 10^{-3}-15 microns	1500 mg/l (1.5%)	>99

Inclined screens cannot remove SS to the same extent as sedimentation tanks. They operate best on trashy materials which may foul subsequent treatment of sludge handling units. Their ability to remove fine grit is limited by size openings. Separate grit removal equipment, if needed, should be installed after the inclined screens.

Rotating wedge wire screens are another design. These tend to overcome grease blinding problems of its own wedge wire screen. They employ a rotating wedge wire screen which backwashes itself. Wastewater passes vertically downward from the outside to the inside of the drum by gravity. The screened wastewater then passes out through the lower half of the drum to a collection trough.

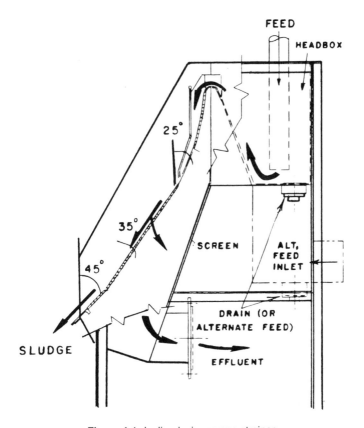

Figure 4-1. Inclined wire screen strainer.

Another major design is a microscreen. The functional design involves:

1. Characterization of suspended solids in feed as to concentration and degree of flocculation. These factors affect microscreen capacity, performance and backwashing requirements.
2. Selection of unit design parameters which assure sufficient capacity to meet maximum hydraulic loadings with critical solids characteristics, and provide the required performance over the expected range of hydraulic loadings and solids characteristics.

3. Provision of backwash and supplemental cleaning facilities to maintain the design capacity.

Table 4-2 gives typical values for microscreen and backwash design parameters for solids removal from secondary effluents. Similar values would apply to direct microscreening of good quality effluent from fixed film biological reactors such as trickling filters or rotating biological contactors, where the microscreens replace secondary settling

Table 4-2
Microscreen Design Parameters

Item	Typical Value
Screen Mesh	20-25 microns
Submergence	75 percent of height 66 percent of area
Hydraulic Loading	5-10 gpm/sq ft of submerged drum surface area
Head-loss (H_L) through Screen	3-6 in
Peripheral Drum Speed	15 fpm at 3 in. (H_L) 125-150 fpm at 6 in. (H_L)
Typical Diameter of Drum	10 ft
Backwash Flow and Pressure	2 percent of throughput at 50 psi 5 percent of throughput at 15 psi

tanks. Microscreening has been used for the removal of algae from un-coagulated lagoon effluents.

The parameters of mesh size, submergence, allowable headloss and drum speed [rpm = peripheral speed/π/4 (diameter)] are sufficient to determine the flow capacity of a microscreen with given suspended solids characteristics.

The filterability index quantifies the effect of the feed solids characteristics on the flow capacity of a particular fabric.[2] It is assumed that at any constant laminar flow rate the headloss, ΔP in ft, through any given strainer fabric will increase exponentially with the volume passed per unit area (V in ft^3/ft^2):

$$\frac{\Delta P}{\Delta P_o} = e^{IV} \qquad (4\text{-}1)$$

In this relation the filterability index is the exponential rate constant I (1/ft).

From the filterability index, a hydraulic capacity relation for continuous operation of a rotating drum microscreen can be expressed as follows[3]:

$$u = \frac{Q}{A} = \frac{\ln\left[\left(\frac{\Delta P}{C_F}\right)\left(\frac{I\phi}{R}\right) + 1\right]}{\left(\frac{I\phi}{R}\right)} \qquad (4\text{-}2)$$

where: u = mean flow through submerged screen area (fps)
Q = total flow through microscreen (cfs)
A = submerged screen area (sq ft)
P = pressure drop across screen (ft)
C_F = fabric resistance coefficient (ft/fps or sec) (clean fabric headloss at 1 fps approach velocity)
I = filterability index (1/ft)
ϕ = decimal fraction of screen area submerged
R = drum rotational speed (rpm)

The expression $\Delta P/C_F$ represents the initial flow velocity through the clean screen as it enters submergence. C_F characterizes the screen fabric, and varies inversely with mesh opening size as follows:

Mesh Size (microns)	Fabric Resistance C_F
	ft/fps
15	3.6
23	1.8
35	1.0
60	0.8

Limits on ΔP reflect screen fabric mechanical strength and anticipated operating conditions.

The relation of parameters in the expression $(I\phi/R)$ shows that the effect of a higher index or faster buildup of headloss on the screen may be offset by maintaining a higher drum rotational speed.

Figure 4-2 is a plot of Q/A vs. $\Delta P/C_F$ for various values of the parameter $I\phi/R$. The plot is given in terms of constant values for the ratio:

$$E = \frac{Q/A}{\Delta P/C_F} \tag{4-3}$$

This is the ratio of the mean velocity through the screen to the initial velocity when the screen enters submergence. $I\phi/R$ values for the ratio E below 0.5 should be used. Above this limit insufficient opportunity is given for a mat to form on the drum and solids removal efficiency is likely to suffer.

The basic screen support structure of a microscreen is a drum-shaped, rigid frame supported on bearings to allow rotation. Designs using water-lubricated axial bearings or greased bearings located on the upper inside surface of the rotating drum allow submergence well above the central axis. Both plastic (polyester) and stainless steel are used for the microscreen media itself. Greater mechanical strength, especially at higher temperatures, is the prime advantage stated for stainless steel. Table 4-3 gives approximate size and power requirements for various microscreen units.

Diatomaceous Earth Filters

These filters produce a high quality effluent but appear unable to handle the solids loadings normally expected in wastewater applications. These employ a thin layer of precoat formed around a porous septum to strain out the suspended solids in the feedwater which passes through the filter cake and septum. The driving force can be by vacuum from the product side or pressure from the feed side. Headloss through the cake increases due to solids deposition until a maximum is reached.

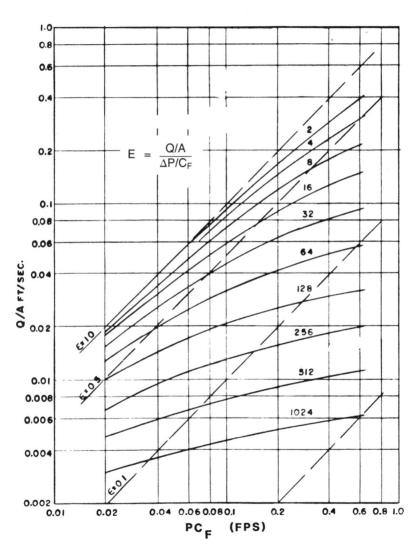

Figure 4-2. Microscreen capacity chart[3].

Table 4-3
Typical Microscreen Power and Space Requirements[1]

Drum Sizes		Floor Space		Motors		Approx. Ranges
Diam.	Length	Width	Length	Drive	Wash Pump	of Capacity
ft	ft	ft	ft	BHP	BHP	mgd
5.0	1.0	8	6	0.50	1.0	0.07 - 0.15
7.5	5.0	11	16	2.00	5.0	0.5 - 1.0
10.0	10.0	14	22	5.00	7.5–5.0	1.5 - 3.0

The cake and associated solids are then removed by flow reversal and the process repeated. In the cases where secondary effluents have been treated, a considerable amount of diatomaceous earth (body feed) is required for continuous feeding with the influent in order to prevent rapid buildup of headlosses. Generally, it is capable of high quality removal of suspended solids but not colloidal matter.

A wide variety of diatomaceous earth (diatomite) grades are available. Coarser grades have greater permeability and solids-holding capacities then do the finer grades which will generally produce a better effluent. Some grades of diatomite are pretreated to change their characteristics for improved performance. A number of vessel configurations are available (open-basin vacuum and vertical pressure designs are most common; see Figure 4-3 for example).

Design criteria are given in References 4-6. The filtration cycle can be divided into two phases, run time and downtime. Downtime includes periods when the dirty cake is dislodged from the septum and removed from the filter and when the new precoat is formed. Run time initiates when the feed is introduced to the filter and ends when a limiting headloss is reached. The body feed rate is the largest operating cost factor. Similarly, it is related to cycle time between backwashing which determines the installed filtering area, hence the capital cost economics.

Ultrafiltration

Ultrafiltration (UF) is a form of membrane separation which employs relatively coarse membrane separation at relatively low pressures[1]. The process should be differentiated from reverse osmosis which is a similar process used for dissolved solids separation using fine membranes

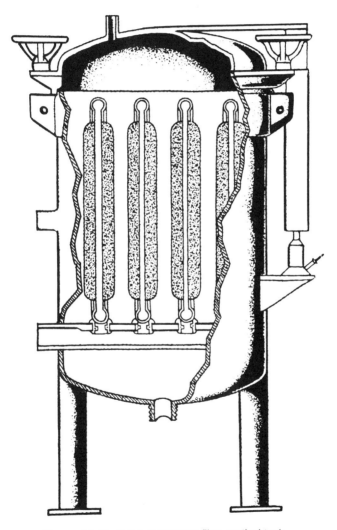

Figure 4-3. Vertical leaf pressure filter, vertical tank.

and high pressures. Ultrafiltration uses a thin semipermeable polymeric membrane. It is most successful in separating suspended solids as well as large-molecule colloidal solids (0.002 to 10.0 μ) from wastewater. Fluid transport and solids retention are achieved by regulating pore size openings. A major drawback of ultrafiltration is the high capital and operating costs. The high cost may be offset by using compactness where space is a critical factor. Important design considerations are: membrane area, membrane configuration, membrane material, membrane life, and driving force.

The required membrane area is a function of flux which is determined by membrane construction and the fouling characteristics of the fluid. By operating the process at liquid velocities of three to ten fps parallel to the membrane surface, scouring of contaminants can be accomplished and a more stable flux achieved. At such velocities, with normal membrane fluxes, single-pass design would require an impractically large membrane area. Wastewater is usually recirculated. A result is some blowdown of concentrated waste to prevent excessive solids buildup.

Membrane configuration concerns the amount of membrane surface area which can be incorporated into a module. Because of low membrane fluxes it is important to design the module to maximize membrane surface area. Common designs include tubular support elements over which a membrane is wound helically or in which the membrane is enclosed in a continuous spiral (see Cheremisinoff et al.[5,6]).

The membrane itself is comprised of two layers: surface—an extremely thin homogenous polymer of 0.1 to 10.00 microns (typically, 5.0 microns), and surface support—an open cell of five to ten mil thickness. The membrane, in turn, is supported on a porous sheet for added mechanical support. The thin surface layer controls the transport and rejection properties of the membrane. Typical membrane specification ranges are given in Table 4-4.

Industrial Equipment

Equipment details and operating principles are described in detail by Cheremisinoff et al.[5,6]. Co-current designs refer to those systems where the direction of gravity force and filtrate motion coincide. Typical systems included in this category are rotary-drum filters, continuous-belt filters, Nutsch batch filters and filter presses with horizontal chambers.

Design features of the internal rotary-drum filter are shown in Figure 4-4. The filter medium is retained on the inner periphery. It is used for rapid settling slurries that do not require a high degree of washing.

Table 4-4
Typical Membrane Specifications for Ultrafilters[1]

Material	Most organic polymers
Water Permeability	7-290 gpd per sq ft at 30 psi
Molecular weight	340-45,000
Retentivity	60-100 percent
Maximum Operating Temperature	50-120°C

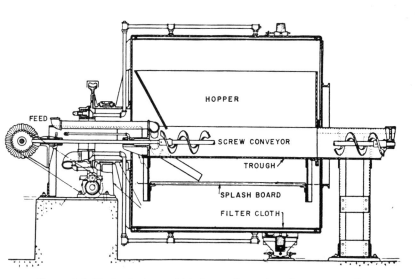

Figure 4-4. Design features of interior-medium rotary-drum vacuum filter.

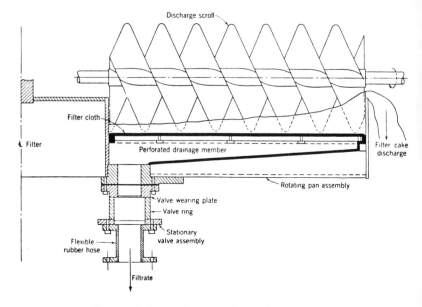

Figure 4-5. Design features of rotary horizontal filter.

Nutsch filters are comprised of a flat filtering plate or a large flat-bottomed tank with a loose filter medium. In vacuum filtration, these false-bottom tanks are the same general design as vessels employed for gravity filtration.

A horizontal rotary filter is shown in Figure 4-5. These are used for filtering quick-draining crystalline solids. The design consists of a circular horizontal table which rotates about a center axis. The table consists of several hollow pie-shaped segments with perforated or woven metal tops. Each section is covered with a filter medium and is connected to a central valve that removes filtrate and wash liquors.

Belt filters consist of a series of Nutsch filters that move along a closed path. A belt filter consists of an endless supporting perforated rubber belt covered with a filter cloth (see Figure 4-6). The velocity of the filtering partition depends on the physical properties of the sludge and the filter length. Cake thickness may range from 1 to 25 mm.

Cross-mode filters have a vertical flat or cylindrical filtering partition. Filtrate moves inside the channels of the filtering elements along the surface of the filter partition downward under gravity force action,

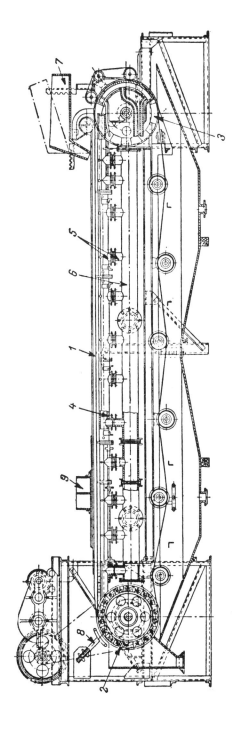

Figure 4-6. Design features of a belt filter: 1. support and filter partition; 2. driving drum; 3. tensioning drum; 4. elongated chamber; 5. nozzles; 6. filtrate collector; 7. sludge feed trough; 8. cake washing device; 9. tank for wash liquor.

Table 4-5
Filtration Ranges for Cartridge Filters[5]

Industry and Liquid	Typical Filtration Range
Petroleum Industry	
Isobutane	250 mesh
MEA	200 mesh to 5–10 μm
Naphtha	25–30 μm
Produced Water for Injection	1–3 to 15–20 μm
Residual Oil	25–50 μm
Seawater	5–10 μm
Steam Injection	5–10 μm
Vacuum Gas Oil	25–75 μm
Pulp and Paper	
Calcium Carbonate	30–100 mesh
Clarified White Water	30–100 mesh
Dye	60–400 mesh
Freshwater	30–200 mesh
Groundwood Decker Recycle	20–60 mesh
Hot Melt Adhesives	40–100 mesh
Latex	40–100 mesh
Millwater	60–100 mesh
Paper Coating	30–250 mesh
River Water	20–400 mesh
Starch Size	20–100 mesh
Titanium Dioxide	100–200 mesh
All Industries	
Adhesives	30–150 mesh
Boiler Feed Water	5–10 μm
Caustic Soda	250 mesh
Chiller Water	200 mesh
City Water	500 mesh to 1–3 μm
Clay Slip (ceramic and china)	20–700 mesh
Coal-Based Synfuel	60 mesh
Condensate	200 mesh to 5–10 μm
Coolant Water	500 mesh
Cooling Tower Water	150–250 mesh
Deionized Water	100–250 mesh
Ethylene Glycol	100 mesh to 1–3 μm
Floor Polish	250 mesh
Glycerine	5–10 μm
Inks	40–150 mesh
Liquid Detergent	40 mesh
Machine Oil	150 mesh
Pelletizer Water	250 mesh
Phenolic Resin Binder	60 mesh
Photographic Chemicals	25–30 μm
Pump Seal Water	200 mesh to 5–10 μm
Quench Water	250 mesh
Resins	30–150 mesh
Scrubber Water	40–100 mesh
Wax	20–200 mesh
Wellwater	60 mesh to 1–3 μm

[a]Courtesy of Ronningen-Petter Division, Dover Corporation, Portage, MI.

or rises along this partition upward under the action of a pressure differential. Common designs are filter presses, leaf filters, and disk filters. These designs are described in detail by Cheremisinoff et al.[5]

Cartridge filters are among the most widely used filtration devices in the chemical and process industries. They range from laboratory-scale to commercial operations for flows extending to an excess of 5000 gpm. Table 4-5 gives typical ranges used in different industries.

Single cartridge units can be piped directly into systems requiring batch or intermittent service. Table 4-6 gives typical particle retention

Table 4-6
Typical Filter Retentions[a]

	Mesh or Mesh Equivalent	Nominal Particle Retention		Percentage of Open Area
		in.	μm	
Wire Mesh	10	0.065	1650	56
	20	0.035	890	46
	30	0.023	585	41
	40	0.015	380	36
	60	0.009	230	27
	80	0.007	180	32
	100	0.0055	140	30
	150	0.0046	115	37
	200	0.0033	84	33
	250	0.0024	60	36
	400	0.0018	45	36
	700	0.0012	30	25
Perforated	10	0.063	1575	15
	20	0.045	1125	18
	30	0.024	600	12
Slotted	10	0.063	1600	50
	15	0.045	1140	43
	20	0.035	890	36
	30	0.024	610	30
	40	0.015	380	20
	60	0.009	230	18
	80	0.007	180	25
	100	0.006	150	13
	120	0.005	125	11
	150	0.004	100	9
	200	0.003	75	7
	325	0.002	50	5
		0.001	25	3
Fabric	60	0.009	230	NA[b]
	80	0.007	180	NA
	100	0.0055	140	NA
	150	0.0046	115	NA
	250	0.0024	60	NA
	500	0.0016	40	NA
		0.0010–0.0012	25–30	NA
		0.0006–0.0008	15–20	NA
		0.0002–0.0004	5–10	NA
		0.00004–0.00012	1–3	NA

[a]Courtesy of Ronningen-Petter Division, Dover Corp., Portage, MI.
[b]NA = percentage of open area not applicable to fabric media.

Table 4-7
Typical Dimensions of Single Filter Units[a]

Body Size (in.)	Inlet/Outlet Diameter NPTI	Filter Element Dimensions (in.)	Filter Element Area (in.^{2})	Straight-Through Filter Dimensions (in.) A	Straight-Through Filter Dimensions (in.) B	Standard Filter Dimensions (in.) C	Standard Filter Dimensions (in.) D
$2^1/_2$	1	2×12	75	$24^1/_{16}$	$19^7/_{16}$	$24^{11}/_{16}$	4
$2^1/_2$	1	2×18	112	$30^1/_{16}$	$25^7/_{16}$	$30^{11}/_{16}$	4
3	$1^1/_2$	$2^1/_4 \times 24$	168	$38^5/_{16}$	$32^{15}/_{16}$	$37^7/_8$	$4^5/_8$
3	$1^1/_2$	$2^1/_4 \times 36$	256	$50^5/_{16}$	$44^{15}/_{16}$	$49^7/_8$	$4^5/_8$
4	2	$3^1/_4 \times 18$	182	$32^7/_8$	$26^1/_2$	$34^{13}/_{16}$	$5^3/_{16}$
4	2	$3^1/_4 \times 36$	364	$51^5/_{16}$	$44^{15}/_{16}$	$52^{13}/_{16}$	$5^3/_{16}$
4	3	$3^1/_4 \times 18$	182.	$33^9/_{16}$	$26^1/_2$		
4	3	$3^1/_4 \times 36$	364	52	$44^{15}/_{16}$		

[a]Courtesy of Ronningen-Petter Division, Dover Corp., Portage, MI.

sizes for different filter media. Typical dimensions are given in Table 4-7. Multiple filters are also common. These consist of two or more single filter units piped in parallel to common headers. Table 4-8 can serve as a rough guide to filter cartridge selection. Sample calculations for sizing these units are given by Cheremisinoff et al.[5]

Filter Media Data

This subsection contains miscellaneous data on the properties and characteristics of commercially available cloth media. This data is useful in making general selection of filter media for a particular application.

Table 4-9 gives general properties of woven filter cloth fibers.

Table 4-10 lists various data on synthetic fabric filters.

Table 4-11 lists physical properties and chemical resistances of polyester fibers for belt filters.

Additional data are given by Cheremisinoff et al.[5,6]

(text continued on page 68)

Table 4-8
Guide to Filter Cartridge Selection[a]

Electroplating Solutions		Filter Tube (Material/Core)
Acid		
Fluoborates	Cu, Fe, Pb, Sn	Polypropylene (PP) or Dynel/PP
Nonfluoborates	Cu, Sn, Zn: <6 oz/gal H_2SO_4	PP or cotton/PP
	Cu, Sn, Zn: >6 oz/gal H_2SO_4	PP or Dynel/PP
	Cr	PP or Dynel/PP
	Au, In, Rh, Pd	PP or Dynel/PP
	$FeCl_2$ (190°F)	PP/rigid PP (RPP) or porous stone
	Ni (Woods)	PP or Dynel/PP
	Ni (Watts type & bright)	PP or cotton/PP
	Ni (high-chloride)	PP or cotton/PP
	Ni (sulfamate)	PP or cotton/PP
	Electrotype Cu and Ni	PP or cotton/PP
Alkaline	Sn (stannate)	Cotton/stainless steel (SS)
Cyanide	Brass, Cd, Cu, Zn[b]	Cotton/SS, PP or Dynel/PP
	Au, In, Pt, Ag	Cotton/SS, PP/PP
Pyrophosphate	Cu, Fe, Sn, etc.	Cotton/SS or PP
Electroless	Ni plating: <140°F	Cotton/SS or PP
	Ni plating: >140°F	PP/RPP, cotton/SS
	Cu: <140°F	PP/PP
	Cu: >140°F	PP/RPP

Chemicals		Filter Tube (Material/Core)
Acids	Acetic: dilute	Cotton/SS, PP/PP
	Acetic: concentrated	PP or Dynel/PP
	Boric	Cotton/SS, PP/PP
	Chromic, hydrochloric, nitric, phosphoric, sulfuric	PP or Dynel/PP, porous stone[c]
	Hydrofluoric, fluoboric	PP or Dynel/PP
Alkalies	NaOH or KOH	PP/PP
	NH_4OH: dilute	cotton/SS, PP/PP
	NH_4OH: concentrated	PP or Dynel/PP
Misc. Chemicals	Biological solutions	Cotton/SS, PP/PP, porous stone[c]
	Electropolishing solutions	Porous stone, PP/PP
	Pharmaceutical solutions	Cotton/SS, PP/PP, porous stone[c]
	Photographic solutions	Cotton/SS, PP/PP
	Radioactive solutions	Cotton/SS, porous stone[c]
	Ultrasonic cleaning solutions	Cotton special B compound/SS
	Nickel acetate (190°F)	Cotton/SS
	Food products	Cotton/SS, PP/PP
Organic Liquids	CCl_4	Cotton/steel or SS
	Dichloroethylene	Cotton/steel or SS
	Hydraulic fluids	Cotton/steel or SS
	Lacquers	Cotton/steel or SS
	Per- and trichloroethylene	Cotton/steel or SS
	Solvents	Cotton/steel or SS
Petroleum Products	Fuel oil, diesel, kerosene, gasoline, lube oil	Cotton/steel or SS

[a]Courtesy of Sethco Division, MET PRO Corp., Hauppauge, NY.
[b]When operated as high-speed baths at high temperatures (>140°F) or with high alkali content, use PP or Dynel/PP.
[c]Porous stone is recommended for all acids except hydrofluoric and fluoboric.

Table 4-9
Properties of Woven Filter Cloth Fibers[5]

Fibers	Acids	Alkalies	Solvents	Fiber Tensile Strength	Temperature Limit (°F)
Acrilan	Good	Good	Good	High	275
Asbestos	Poor	Poor	Poor	Low	750
Cotton	Poor	Fair	Good	High	300
Dacron	Fair	Fair	Fair	High	350
Dynel	Good	Good	Good	Fair	200
Glass	High	Fair	Fair	High	600
Nylon	Fair	Good	Good	High	300
Orlon	Good	Fair	Good	High	275
Saran	Good	Good	Good	High	240
Teflon	High	High	High	Fair	180
Wool	Fair	Poor	Fair	Low	300

Table 4-10
Technical Data for Various Synthetic Fabric Filters

Style No.	Weave	Weight (oz/yd²)	Threads/in., Warp × Weft	Thread Diam., Warp × Weft (μm)	Mesh Opening (μm)	Air Permeability (ft³/min)	Thickness (μm)
Nylon 6,6,6							
Warp and Weft Monofilament							
111-020	Plain	4.5	22 × 22	305 × 305	850 × 850	NA[b]	570
111-110	Plain	4.6	50 × 50	200 × 200	300 × 300	NA	350
111-150	Plain	3.2	62 × 62	150 × 150	250 × 250	NA	270
111-160	Plain	2.4	107 × 76	100 × 100	250 × 230	NA	220
111-170	Plain	4.6	29 × 29	250 × 250	600 × 600	NA	450
111-180	Plain	2.3	66 × 66	130 × 130	210 × 210	NA	270
111-190	Twill	5.5	38 × 38	250 × 250	420 × 420	NA	530
111-206	Plain	5.3	183 × 43	150 × 150		170-210	410
111-220	Plain	2.9	80 × 80	125 × 125	175 × 175	NA	240
111-230	Plain	5.7	40 × 40	250 × 250	420 × 420	NA	450
111-056	Satin	7.2	109 × 42	205 × 300		350-400	450
Warp and Weft Multifilament							
1053	Plain	2.1	147 × 97			150-200	160
1093	Twill	3.5	297 × 122			75-100	250
1103	Twill	3.7	297 × 135			40-70	250
1123	Twill	3.2	195 × 140			15-25	190
1153	Satin	3.2	236 × 99			15-25	210
1193	Satin	5.3	300 × 99			45-70	300
1203	Satin	6.8	152 × 76			20-30	390
1212	Satin	6.5	178 × 97			50-80	710
1283	Plain	1.8	112 × 97			170-220	130
1338	Leno	5.6	7 × 5			NA	630
1353	Plain	3.2	64 × 48			50-100	255
1363	Plain	11.5	69 × 28			1-3	660
1393	Plain	9.7	236 × 53			5-10	560
122-053	Twill	4.7	80 × 117			50-80	410
122-073	Oxford	15.5	72 × 21			0.5-2	720

(continued on next page)

(Table 4-10 continued)

Style No.	Weave	Weight (oz/yd²)	Threads/in., Warp × Weft	Thread Diam., Warp × Weft (µm)	Mesh Opening (µm)	Air Permeability (ft³/min)	Thickness (µm)
Warp Monofilament, Weft Spun							
1233	Satin	5.0	320 × 71			60–100	410
Warp and Weft Multifilament and Metal Spun							
9165	Twill	4.1	297 × 132			25–40	300
Nylon 11							
Warp and Weft Monofilament							
1656	Satin	8.2	99 × 53	180 × 290		200–300	520
1666	Satin	9.0	111 × 53	180 × 290		125–200	530
1686	Satin	7.1	107 × 91	180 × 180		125–200	395
111–096	Satin	9.5	99 × 53	205 × 300		150–300	490
Nomex							
Warp Multifilament, Weft Spun							
1513	Plain	7.7	107 × 66			50–80	510
Polyester							
Warp and Weft Monofilament							
1713	Plain	4.1	345 × 84			80–120	190
1716	Plain	4.1	350 × 79			10–20	180
1733	Plain	8.8	175 × 36			300–400	380
9656 B	Satin	8.3	104 × 53	200 × 300		300–400	450
9813 B	Satin	11.8	112 × 53	200 × 300		200–250	510
9884 B	Satin	12.7	145 × 53	200 × 250		80–150	440
311–010 HS[c]	Plain	6.9	38 × 38	250 × 250	420 × 420	NA	400
311–020	Plain	6.9	38 × 38	250 × 250	420 × 420	NA	400
Warp and Weft Multifilament							
1813	Plain	10.3	43 × 30			1–3	500
1853	Leno	10.0	7 × 5			NA	800
1893	Twill	4.1	295 × 112			40–75	250
1896	Twill	4.1	295 × 112			5–10	200
1933	Satin	3.2	290 × 99			35–55	190
1943	Twill	6.5	257 × 79			10–20	330
1953	Plain	10.0	36 × 30			2–4	420
1956	Plain	10.0	36 × 30			1–2	380
1973	Plain	3.5	94 × 81			20–40	195
1976	Plain	3.5	99 × 84			5–15	140

Table 4-11
Physical Properties and Chemical Resistances of
Polyester Fibers Used for Belt Filters[a]

Specific Gravity	1.38
Moisture Regain	
At 65% RH and 68°F (20°C) (%)	0.4
Water Retention Power (%)	3–5
Tensile Strength	
cN/dtex	7–9.5
Wet in % of dry	95–100
Elongation at Break	
%	10–20
Wet in % of dry	100–105
Ultraviolet Light Resistance	R[b]
Resistance to Fungus, Rot and Mildew	R
Resistance to Dry Heat	
Continuous	
°F	302
°C	150
Short-Term Exposure	
°F	392
°C	200
Chemical Resistance to	
Acids	C[c]
Acetic Acid Concentration	R
Sulfuric Acid 20%	R
Nitric Acid 10%	C
Hydrochloric Acid 25%	C
Alkalies	C
Saturated Sodium Carbonate	R
Chlorine Bleach Concentration	R
Caustic Soda 25%	U[d]
Ammonia Concentration	U
Potassium Permanganate 50%	R
Formaldehyde Concentration	R
Chlorinated Hydrocarbons	R
Benzene	R
Phenol	C
Ketones, Acetone	R

[a]Courtesy of Industrial Fabrics Corp., Minneapolis, MN.
[b]R = recommended.
[c]C = conditional.
[d]U = unsatisfactory.

(text continued from page 62)

References

1. *Process Design Manual for Suspended Solids Removal,* U.S. Environmental Protection Agency, EPA 625/1-75-003a (Jan. 1975).
2. Boucher, P. L., *ICE J.* (Brit.), 24, 4, p. 415 (1947).
3. Mixon, F. O., *J. WPCF,* 42, p. 1944 (Nov. 1970).
4. Bell, G. R., *J. AWWA,* 54, p. 1241 (Oct. 1962).
5. Cheremisinoff, N. P. and D. Azbel, *Liquid Filtration,* Ann Arbor Science Pub., Ann Arbor, MI (1983).
6. Cheremisinoff, P. N. and R. A. Young, *Pollution Engineering Practice Handbook,* Ann Arbor Science Pub., Ann Arbor, MI (1975).

5

DEEP BED FILTRATION

General Description

Filtration is accomplished by passing the fluid through a bed composed of granular material with or without the addition of chemicals. The end of the filter run (filtration phase) is reached when the suspended solids in the effluent start to increase (breakthrough) beyond an acceptable level, or when a limiting head loss occurs across the filter bed. In filters of this type, there are two primary means whereby particles are removed from the stream: *entrapment* and *adhesion*.

The size of pore channels determines the size of particles that can be removed by mechanical means. If the particles are granular in nature, there is a certain size above which 100% filtration efficiency will be achieved. However, most applications deal with particles (e.g., oil, silt, floc) which will deform under the hydraulic forces to which they are subjected as they pass through the media bed.

For sand, the void areas are more uniform than for anthracite or garnet. This means that a finer filtration can be achieved with a sand bed; however, it also results in lower solids loadings per sq ft and higher pressure losses per run times. Also, the large particles are removed in the upper portion of the media bed filling up the available void areas and further shortening filtration cycles.

Graded media beds composed of coarse and fine particles are desirable. This cannot be achieved with beds using only one media (e.g., sand). The coarser material will have a greater specific gravity and the bed will reclassify itself in a fine to coarse gradation during the backwash fluidization. Therefore, media of different types are employed. Typically, this dual media configuration consists of a combination of a layer of anthracite over a layer of sand or garnet. The specific gravity

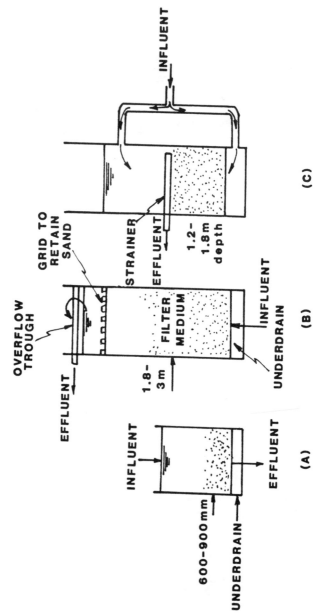

Figure 5-1. Types of filters used: (A) Conventional; (B) Upflow; (C) Biflow.

of anthracite is approximately half that of sand, and the garnet is even heavier than sand.

Conventional designs employ a coarser media layer on top to remove the larger solids and/or flocs. A finer "polishing" layer is located beneath and is able to maintain the gradation after backwash. By using this type of dual media configuration, economical filtration cycles can be maintained even on relatively high TSS (total suspended solids) waters (100–150 ppm).

All filters may be classified according to the direction of flow as downflow, upflow, or biflow filters. See Figure 5-1. The downflow filter is the most commonly used.

In upflow filters, the flow passes up through the granular medium. The medium stratifies after backwashing, with coarse material on the bottom and fines at the top. The upflow filter is generally more efficient because the passage of the filterable liquid in the upward direction allows greater penetration of the suspended solids into the bed. During filtering, the granular medium is retained by a metal grid located at the top of the filter. The filtering medium forms an inverted arch with the evenly spaced bars. If the arch structure is disrupted by increasing the flowrate through the filter or by injecting air into the filter, the medium moves up past the retaining bars. This principle is used in the backwashing or the upflow filter. Liquid being filtered can also be used for backwashing.

The combination of both the downflow and the upflow filters are combined in the biflow design. Here, effluent is collected through a strainer placed within the bed. Backwashing is accomplished by increasing the flowrate to the bottom of the filter. A granular media filter is intended to filter in depth. Solids removal takes place within the filter, and not primarily at the entering surface.

A number of filter configurations have been developed to accommodate the higher solids loads and to encourage filtration in depth. The alternates were developed because the conventional sand filter is not well-suited to handle high influence solids loads. This is due to the fact that the media is backwashed at a rate sufficient to expand the bed, and stratification of the media occurs—the smaller sand grains tending to collect near the top of the bed and the larger grains at the bottom. Since the conventional sand filter is operated with downward flow, the solids encounter the finest media first, and filtration in depth is not achieved. The newer configurations achieve longer filter cycles with higher influent solids loads. The most common method is the double-layer bed composed of a coarse layer of crushed anthracite coal over a layer of finer sand. Coal has a lighter specific gravity than sand, and when the proper size, it remains on top during backwashing if the backwash rate is sufficient to achieve full fluidization and minimal expansion of the bed. The coarse coal removes the major portion of the sediment, allowing significant depth of penetration, while the fine sand layer polishes the water to provide an effluent of good quality. Crushed coal is angular in shape and has a higher porosity than sand (about 0.50 vs. 0.40). The coal has a greater storage capacity for solids removed. An additional layer of finer and heavier garnet under the silica sand can also be used. The three layers are sized to encourage some intermixing between layers.

Practice also includes upflow and middle outlet sand filters. The upflow filter achieves its benefit because of the graded gravel support at the bottom which acts as a roughing filter prior to the deep bed of unstratified coarse sand above. There is a danger of lifting part of the filter bed during the filtration cycle. This type of filter has received considerable attention in Great Britain for wastewater filtration because of its high solids capacity. The lifting problem is overcome by the use of an inlet at both top and bottom in biflow designs. The flow automatically distributes itself so that the upward and downward forces balance each other, eliminating the uplift danger of the upflow filter.

The advantages of the dual media filter include the fact that the practice is well established in potable water practice, and the increased solids handling capacity of the coal layer is achieved with a relatively shallow media because the sand layer is provided to protect the filtrate quality. The shallow media reduces structural costs. One disadvantage is that full fluidization is essential during the water backwash to ensure that the coal will remain on top after backwash. This requirement may dictate excessive wash rates if a coarse coal is selected to lengthen filter cycles. Another disadvantage is incurred if media-retaining underdrain strainers are provided to eliminate the use of supporting graded gravel. The fine sand of the lower layer requires very fine openings in the strainers, and clogging problems may result.

The triple media filter has similar requirements, advantages, and disadvantages. These filters are usually supported on graded gravel, this to avoid the dangers of strainers with fine openings.

The advantages of the single media unstratified filter are:

1. The simplicity of a single media.
2. The ready availability of the media.
3. Coarser media may be used to achieve longer filter cycles without excessive backwash rates since these filters are backwashed with

air and water simultaneously without bed expansion (i.e., below the water fluidization velocity).

4. Larger underdrain strainer openings may be used and thus, less danger of strainer clogging.

5. Transport of heavy solids released from the filter during backwash to the overflow is more effective with simultaneous air and water wash than with water alone.

The disadvantages of the single media unstratified filter are:

1. A deeper bed is needed to try to compensate for the coarser media resulting in higher structural costs.

2. Deeper beds do not necessarily ensure equivalent filtrate quality compared with a dual or triple media bed.

3. Precautions are necessary to ensure that filter media is not lost during the air-plus-water backwashing operation.

The advantages of the upflow filter are:

1. Unfiltered water is usually used for backwashing without danger of strainer clogging since strainers are not used in the underdrain.

2. There is no need to provide substantial water depth above the media for head development as is desired in downflow gravity filters to avoid negative head development.

3. Low backwash water rates are used because part of the wash cycle is with air and water simultaneously.

The disadvantages of the upflow filter include:

1. The total bed depth is large adding to the structural costs.

2. Media loss during the air and water backwash has exposed the hold-down grid in some plants, leading to uplift difficulties at high headlosses.

3. The backwash routine including draindown is fairly long, about 45–60 minutes depending on the details.

4. The initial filtrate after backwash can be no better than the backwash supply quality which can be quite poor at times.

Solids Loading

As more and more material is removed from the stream, the pore sizes between the media grains decrease. This increases the pressure drop across the bed and increases the velocity of the liquid through the remaining flow channels. These increased hydraulic forces overcome some of the adhesive forces holding the particles thus forcing the particles deeper into the bed where the velocities and the specific pressure drop are not as great. Here the particles will reattach themselves until the loading is such that the process repeats itself.

The velocity of the liquid in the open area above the media can be calculated from:

$$\frac{\text{Gal/Min}}{\text{Sq Ft}} \times \frac{0.1337 \text{ Cu Ft}}{1 \text{ Gal}} = \frac{\text{Ft}}{\text{Min}} \times \frac{12 \text{ In.}}{1 \text{ Ft}} = \frac{\text{In.}}{\text{Min}} \qquad (5\text{-}1)$$

Media with 0.3 mm effective sizes has a void area of approximately 55% so the velocity in the media bed itself increases to 35 in./min or 0.58 in./sec. This velocity is extremely low when compared to the average velocity through the process piping, typically at 7–12 fps (85–144 in./sec). Although there is a further increase in velocity as some of these void areas are decreased or even bridged during filtration, the velocity still remains low enough for adhesion to be an effective filtration mechanism. This is especially true because below the area of solids built up, the velocity will decrease once again to the 35 in./min rate. This line of solids build up and increased velocity moves down through the bed as a solids loading front during filtration.

The solids loading per sq ft is also affected by the use of dual media. The dual media configurations provide two-stage removal of contaminate particles. Let's take a closer look at how this affects solids loading. As noted, waste waters have particles of various sizes. These sizes tend to fall mostly in one of two groups—one in the range of 1–5 microns and one in the 50–150 micron range. In any media filter, the majority of the contaminates are removed in the uppermost part of the media layer. The further down in the bed samples are taken, the fewer number of particles that are retained. However, in a dual media filter, there are effectively two media surfaces, the anthracite surface and the sand surface. If we were to take case samples and analyze them for the amounts of contaminates present per inch, the performance of a single versus a dual media filter would resemble that of Figure 5-2. This illustrates that higher solids/sq ft loadings can be achieved with dual media filters while still maintaining the polishing effectiveness of a single media sand filter. Thus dual media with proper polymer additions when necessary is a highly effective means of removing particles even down to 1–10 microns for many applications.

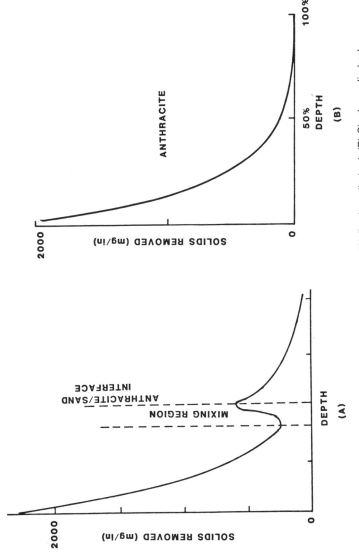

Figure 5-2. Solids removal as a function of depth: (A) Dual media bed; (B) Single media bed.

Design Considerations

The filtration driving force is when the force of gravity or an applied pressure force can be used to overcome the frictional resistance to flow offered by the filter bed. Gravity filters are commonly used for the filtration of treated effluent at large utility plants. Pressure filters operate in the same manner as gravity filters and are used at smaller plants. The only difference is that, in pressure filters, the filtration operation is carried out in a closed vessel under pressurized conditions achieved by pumping. Pressure filters normally are operated at higher terminal head losses. This generally results in longer filter runs and reduced backwash requirements.

Typical filtration schemes for wastewater treatment are shown in Figure 5-3. Objectives of these designs are:

1. Removal of residual biological floc in settled effluents from secondary treatment by trickling filters or activated-sludge processes. This is referred to as "tertiary filtration."
2. Removal of precipitates resulting from alum, iron, or lime precipitation of phosphates in secondary effluents from trickling filters or activated-sludge processes. The suspended solids to be filtered can be substantially different from those in normal secondary effluent.
3. Removal of solids remaining after the chemical coagulation of wastewaters in physical-chemical waste-treatment processes, i.e., following lime treatment of raw wastewater and before adsorption removal of soluble organics in carbon columns. Again, the solids to be filtered can be substantially different from normal secondary effluent solids.

Filters can be used as the final process of wastewater treatment (polishing secondary or tertiary effluents) or as an intermediate process to prepare wastewater for further treatment (for example, before downflow carbon adsorption columns). The required filters should be designed to provide a quality of filtrate equal to or better than the desired effluent-quality goal. Achieving this quality may require a pilot study to evaluate the flow characteristics and solids characteristics of the water to be filtered. In the absence of a pilot study, the design must be based on experience with similar filter influent waters at other installations.

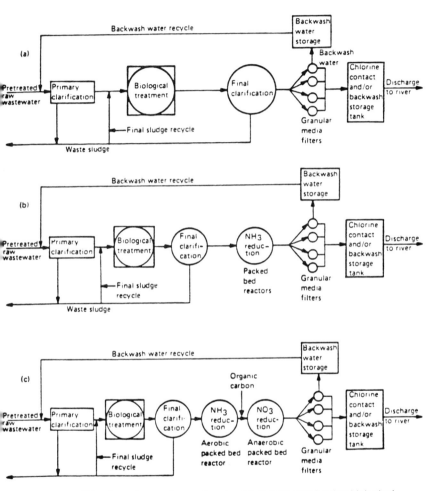

Figure 5-3. Filtration schemes for wastewater treatment: (a) following biological secondary treatment for carbonaceous BOD removal; (b) following biological secondary and biological tertiary (packed-bed reactors) treatment for carbonaceous BOD and ammonia reduction; (c) following biological secondary and biological tertiary (packed-bed reactors), both aerobic and anaerobic for carbonaceous BOD, ammonia and nitrate reduction. (Phosphorous levels may also be reduced by adding ferric and aluminum salts and a polymer feed to solids contact units located ahead of the granular-media filters.)

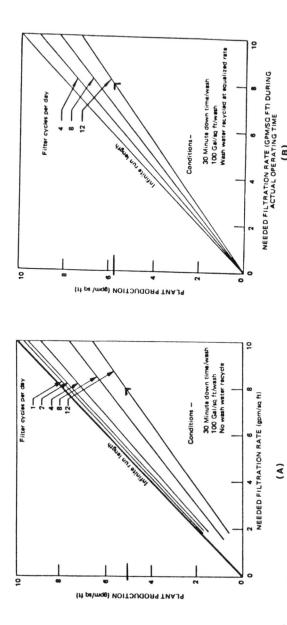

Figure 5-4. (A) Effect of number of filter cycles per day on filtrate production when using filtered water for backwashing; (B) effect of number of filter cycles per day on filtrate production when using unfiltered water for backwashing. Plant production is the average plant output over the full 24-hour day.

The problem of potential short filter cycles during peak-load periods requires that a flow equalization tank be considered as part of the flow scheme. It may be installed near the plant entrance to benefit the entire treatment operation, or installed just prior to the filters. Figure 5-3 shows that at this point, wastewater quality is such as to present no odor or mixing problems. Fifteen to twenty percent of mean daily-flow storage capacity would permit constant-rate flow to the filters for a 24-hour period. One hundred percent of mean daily-flow storage capacity would permit constant flow and nearly constant solids loads to the filters.

The cost of a filter is a function of the area of filter provided; hence, the use of a high filtration rate is usually preferred. Filter design should maximize the net water production per sq ft of filter consistent with filter operating feasibility. There are two alternatives. The first occurs when filtered water is used for backwashing. The second case occurs when unfiltered water is used in backwashing. Both cases are illustrated by the plots in Figure 5-4.

The data for both figures were calculated assuming 30 minutes total downtime per backwash to allow for draindown time, auxiliary scour time, actual backwash time and startup time to reach normal rate. The 100 gal/ft^2 total wash water per backwash is typical of volumes adequate for most filtration situations. In the case of recovered wash water, it is assumed that dirty wash water would pass through a holding tank to permit flow equalization of the recirculated water.

Some filters require more than 30 minutes to complete a backwash cycle, especially if complete gravity draindown is essential or desired. Some require more than 100 gal/ft^2/wash. If the downtime and water use for a particular type of filter are expected to deviate significantly from those used above, then the figures should be revised and the cycle length decision reconsidered.

Selection criteria for size and depth of filter media and the appropriate filtration rate are interrelated. Filtrate quality improves with finer media, greater media depth or lower filtration rates. With some influent suspensions, these generalizations are not demonstrated significantly. In filtration of secondary effluents, filtration rate has little effect upon filtrate quality over the usual range of rates employed—two to five gal/min/ft^2—and increased media depth may not compensate for coarser media in achieving filtrate quality.

Selection of filter media also determines the required backwash. Backwash requirements become an integral part of the media decision.

Granular filter media commonly used in water and wastewater filtration include silica sand, garnet sand, and anthracite coal. These media can be purchased in a broad range of effective sizes and uniformity coefficients. The term "effective size" indicates the size of grain (in millimeters) such that 10% (by weight) of the particles are smaller and 90% larger than itself. "Uniformity coefficient" designates the ratio of the size of grain which has 60% of the sample finer than itself, to the effective size which has 10% finer than itself. The media have specific gravities approximately as follows:

1. Anthracite coal, 1.35 to 1.75; most U.S. anthracite, 1.6 to 1.75; U.K. anthracite, 1.35 to 1.45.
2. Silica sand, 2.65.
3. Garnet sand, 4 to 4.2.

The effects of the strong surface removal tendency previously discussed for wastewater filtration must be counteracted by selecting a media size where the flow enters the media which will ensure that the bulk of the suspended solids removal does not occur at the entering surface. The following guidelines should be considered in media size selection:

1. For the tertiary filtration of secondary effluents, media size of at least 1.2 mm is required, and coarser media is preferred if appropriate backwash is provided. Benefits to filter run length accrue at least up to 2.3 mm.
2. For the filtration of chemically treated secondary effluents, a media size of not less than 1.0 mm has been suggested[2].
3. Once the size of the media at the entering surface has been selected, the balance of the media specification is dependent thereon. The uniformity coefficients, the size of the sand in dual media, and the depth of each media must be selected.
4. Low uniformity coefficients (UC) are desired to ensure easier backwashing. This is especially true where fluidization of the media is required during backwashing as with dual and triple media filters. For dual and triple media this is important because the entire media should be fluidized to achieve restratification. The greater the UC (i.e., less uniform size range), the larger the backwash rate required to fluidize the coarser grains. A UC of less than 1.3 is not generally practical because of the sieving capabilities of commercial suppliers. A UC of less than 1.5 can be obtained at a cost premium.
5. A UC of less than 1.5 has the advantage that it will ensure that the coarser grain size in the media (such as the 90% finer size, d_{90}) is not excessively large, requiring a large backwash rate.

6. An alternate method of specifying filter media is to specify the range of size within which the media must fall. A 1.4 to 2.4 mm size range would fall between a U.S. standard 14-mesh and 8-mesh sieve. Some tolerance must be allowed at either end to allow for the sieving capabilities of the suppliers. A 10% tolerance at each end is suggested, 10% by weight could be smaller than 1.4 mm and 10% coarser than 2.4 mm. This system of specification has the advantages that the effective size could be no smaller than the lower end of the range, and the coarser media is more precisely limited which is of importance in selecting the needed backwash rate.

A brief description of the principle types of media follows:

Filter sand consists of hard, durable grains of silica, either irregular or rounded, free from dirt, clay, or organic matter, and should not contain more than 1% of flat or micaeous particles.

Physical characteristics are:

Specific gravity:	2.7
Density (approx.)	100 lbs/cu ft
Calcium and magnesium	$1\frac{1}{2}\%$ by weight as $CaCO_3$ max.
Acid solubility	5% by weight in warm HCl/24-hrs max.
Loss on ignition	2% by weight max.
Effective size	0.45 to 0.55 mm
Uniformity Coefficient	1.25 to 1.40

Support gravel should be round, hard river-run gravel. It should be washed and graded and consist of mainly silica with less than 1% dirt, clay and organic material and less than 3% flat or oblong pieces. Crushed rock is not acceptable. Its physical characteristics are similar to sand and commercial size ranges are:

Fine Gravel	$\frac{1}{8}'' - \frac{1}{4}''$
Medium Gravel	$\frac{3}{8}'' - \frac{3}{4}''$
Coarse Gravel	$\frac{3}{4}'' - 1\frac{1}{2}''$

Anthracite should be crushed anthracite coal, washed and graded. The anthracite should be as free as commercially possible of clay, dirt, shale, ash, and iron compounds. Physical characteristics are:

Specific gravity	1.5–1.6
Density	52.5 lbs/cu ft

Table 5-1
Comparison of Openings to Sieve Designation

Inches	Millimeters	Tyler Mesh	U.S. Std. No.
.093	2.362	8	8
.078	1.981	9	10
.065	1.651	10	12
.055	1.397	12	14
.046	1.168	14	16
.039	.991	16	18
.032	.833	20	20
.027	.701	24	25
.023	.589	28	30
.019	.495	32	35
.016	.417	35	40
.013	.351	42	45
.011	.295	48	50
.0097	.246	60	60
.0082	.208	65	70
.0069	.175	80	80
.0058	.147	100	100

Acid solubility	Negligible in 40% HCl
Alkaline solubility	Less than 3% by weight
Hardness, Mohs Scale	In 1% NaOH at 190°F/1 hr
Effective size	0.9–1.1 mm
Uniformity coefficient	1.85 maximum

The effective size of any media is the size of the screen which would pass only 10% of the material. In other words, 90% of the particles are larger than the effective size of the media (generally expressed in millimeters).

The uniformity coefficient is derived by taking the ratio of the screen size which would pass 60% of the media to the screen size which would pass 10% of the material (the effective size) or:

$$\text{Uniformity Coefficient} = \frac{\text{Screen Size which passes 60\% (mm)}}{\text{Screen Size which passes 10\% (mm)}} \quad (5\text{-}2)$$

The closer UC is to unity the more uniform the media. Effective size and uniformity coefficients are found by plotting the sieve analysis which is the cumulate percentage passed or retained on various sieves vs. the sieve openings. Note that there are two standard sieves listed.

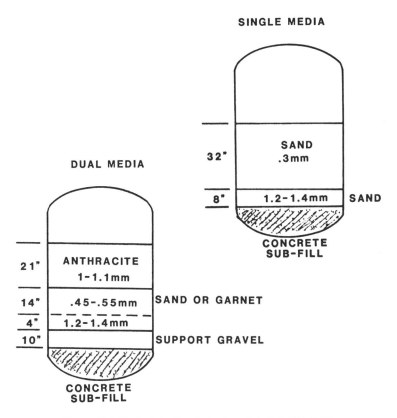

Figure 5-5. Typical depths of single and dual media filters.

Tyler and United States Standard type. Both are acceptable for media sizing (refer to Table 5-1).

The size and type of medias can vary with the application. Typical sizes and depths are shown in Figure 5-5. These are the standards used and represent those which cover the widest range of applications, but are not the only media configurations which can be supplied. Pilot tests are required to determine the optimum media configuration to provide maximum solids removal and filter run time.

References

1. Cheremisinoff, N. P. and D. S. Azbel, *Liquid Filtration,* Ann Arbor Science Pub., Ann Arbor, MI (1983).
2. "Process Design Manual for Upgrading Existing Wastewater Treatment Plants," U.S. Environmental Protection Agency, Washington, DC (Oct. 1974).

6

GRAVITY SEDIMENTATION

Thickeners and Clarifiers

Gravity separation refers to the removal of SS whose specific gravity difference from that of water causes them to settle (or rise) during flow though a basin under quiescent conditions. Sedimentation is divided into *thickening* (increasing the concentration of the feed stream) and *clarification* (the removal of solids from a relatively dilute stream).

In a batch operating mode a thickener normally consists of a standard vessel filled with a suspension. After settling, the clear liquid is decanted and the sediment periodically removed.

Figure 6-1 shows a cross section view of a standard continuous thickener. The drive powers a rotating rake mechanism. Feed enters the apparatus through a feedwell designed to dissipate the velocity and stabilize the density currents of the incoming stream. Separation occurs when the heavy particles settle to the bottom of the tank. Some processes add flocculants to the feed stream to enhance particle agglomeration to promote faster or more effective settling. The clarified liquid overflows the tank and is sent to the next stage of a process. The underflow solids are withdrawn from an underflow cone by gravity discharge or pumping. Cheremisinoff et al.[1,2] provide further descriptions and typical size ranges of different configuration thickeners.

Thickeners can be operated in a countercurrent fashion. Applications are aimed at the recovery of soluble material from settleable solids by means of continuous countercurrent decantation (CCD). The basic scheme involves streams of liquid and thickened sludge moving countercurrently through a series of thickeners. The thickened stream of solids is depleted of soluble constituents as the solution becomes enriched.

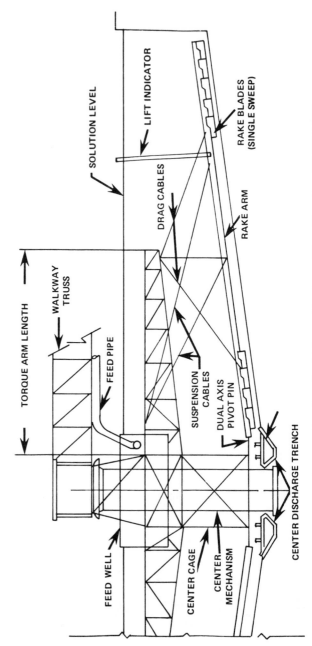

Figure 6-1. Cross section view of a thickener. (Courtesy of Dorr-Oliver Inc., Stamford CT.)

In each successive stage a concentrated slurry is mixed with a solution containing fewer solubles than the liquor in the slurry, and then fed to the thickener. As the solids settle, they are removed and sent to the next stage. The overflow solution, which is richer in the soluble constituent, is sent to the preceding unit. Solids are charged to the system in the first-stage thickener, from which the final concentrated solution is withdrawn. Wash water or virgin solution is added to the last stage, and washed solids are removed in the underflow of this thickener.

Continuous clarifiers handle a variety of process wastes, domestic sewage, and other dilute suspensions. They resemble thickeners in that they are sedimentation tanks or basins whose sludge removal is controlled by a mechanical sludge-raking mechanism. They differ from thickeners in that the amount of solids and weight of thickened sludge are considerably lower. Figure 6-2 shows various types of clarifier configurations. Most cylindrical units are equipped with peripheral weirs; however, some designs include radial weirs to reduce the exit velocity and minimize weir loadings.

Sedimentation Theory

Characteristics of sedimentation are best understood through batch settling experiments as in Figure 6-3. Narrow size range particles settle with about the same velocity. The demarcation line is observed between the supernatant clear liquid (zone A) and the slurry (zone B).

Particles near the bottom of the cylinder pile up, forming a concentrated sludge (zone D), whose height increases as the particles settle from zone B. As the upper interface approaches the sludge buildup on the bottom of the container, the slurry appears more uniform as a heavy sludge (zone D), the settling zone B disappears and the process from then on consists only of the continuation of the slow compaction of the solids in zone D.

Measurement of the interface height and solids concentrations in the dilute and concentrated suspensions can be presented graphically [Figure 6-3(B)]. The plot shows the difference in interface height plotted against time which is proportional to the rate of settling and concentration.

Figure 6-4 shows a plot of sediment height z versus time, t. The sedimentation rate of the heavy sludge decreases with time, which corresponds to the curve on the graph after point K.

The solids concentration in the dilute phase is constant up to the point of complete disappearance of phase A. This is illustrated by the plot in

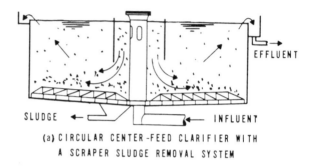

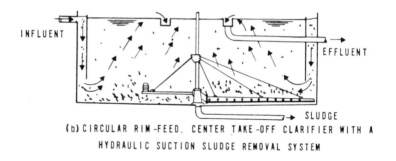

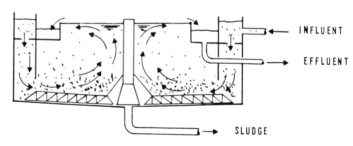

Figure 6-2. Typical clarifier configurations.

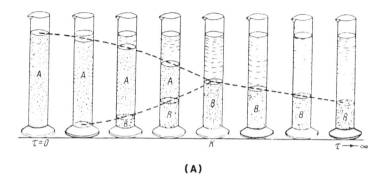

(A)

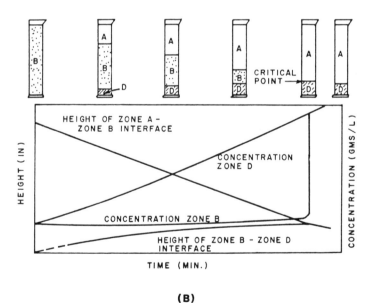

(B)

Figure 6-3. (A) Batch sedimentation in a glass cylinder. (B) Plots of interface height and solids concentration vs. time for batch settling.

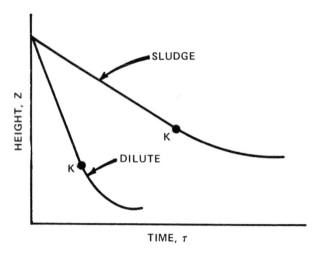

Figure 6-4. Plot describing the kinetics of sedimentation.

Figure 6-5, and corresponds to a constant rate of sedimentation. The concentration in phase B changes with height z and time, t (refer to Figure 6-4), and hence, each curve in Figure 6-5 represents the distribution of concentrations at any given moment. The initial concentration is C_1, which remains in the dilute phase during the process. After a sufficient period of time, the concentration increases to C_2, but in zone D. Obviously, if the concentration of the feed suspension is too high, no dilute phase will exist even during the initial period of sedimentation. In this case, concentration and not height will change with time.

Void fraction is defined as the ratio of the liquid volume V_f filling the space among the particles, to the total volume (the sum of the liquid volume and the actual volume of the solid particles V_p):

$$\epsilon = \frac{V_f}{V_p + V_f} \qquad (6\text{-}1)$$

As particles settle, forming a thickened zone, the void fraction ϵ decreases. Upon total settling of the slurry the void fraction is at a minimum; the value of which depends on the shape of the particles. For example, the minimum void fraction of spherical particles is ϵ_{min} = 0.215; for small crystals, ϵ_{min} = 0.4. For most systems, the void fraction of a thickened sludge is approximately $\epsilon_{min} \simeq 0.6$, however, values should be experimentally determined for the system.

As the sludge compacts, its void fraction and height X decrease. If the initial void fraction ϵ_0 is known when the sludge height is X_0, an average void fraction ϵ can be estimated assuming that the height of the sludge decreases to x. For a vertical cylinder of cross-sectioned area A, the initial volume of the sediment is AX_0, and hence, the volume of solid particles in the sediment is $(1 - \epsilon_0)X_0A$. Similarly, the volume of solid particles for void ϵ is $(1 - \epsilon)XA$. Consequently:

$$\epsilon = \frac{X_0}{X}(1 - \epsilon_0) \tag{6-2}$$

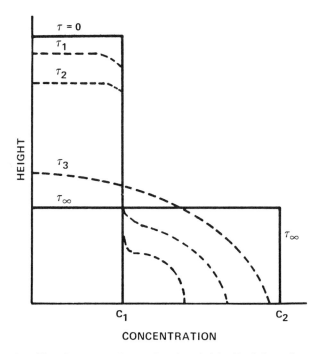

Figure 6-5. Plot of concentration vs. interface height. Dash lines denote sedimentation rates.

For a unit volume of slurry, its weight γ is the sum of the weights of the solid particles $\gamma_p(1 - \epsilon)$ and of the liquid $\gamma_f\epsilon$ (where γ_f = specific weight of liquid; γ_p = specific weight of particles):

$$\epsilon = \frac{\gamma}{\gamma_p - \gamma_f} \tag{6-3}$$

Equation 6-3 can be used to compute the void fraction from experimentally determined values of the specific weights.

The settling velocity for discrete spherical particles is:

$$u = \left[\frac{d^2(\gamma_p - \gamma_f)}{18\mu}\right] \epsilon^2\Phi(\epsilon) \tag{6-4}$$

where $\Phi(\epsilon)$ denotes a function describing the void fraction.

The term in parentheses expresses the velocity of free falling according to the Stoke's law:

$$u_p = \frac{d^2(\gamma_p - \gamma_f)}{18\mu} \tag{6-5}$$

or

$$u = u_p\epsilon^2\Phi(\epsilon) \tag{6-6}$$

For very dilute suspensions (i.e., $\epsilon = 1$ and $\Phi(\epsilon) = 1$), the settling velocity equals the free fall velocity.

Since no valid theoretical expression for the function $\Phi(\epsilon)$ is available, experimental data must be used. Note that a unit volume of thickened sludge contains ϵ volume of liquid and $(1 - \epsilon)$ volume of solid phase (i.e., a unit volume of particles of sludge contains $\epsilon/(1 - \epsilon)$ volume of liquid). Denoting σ as the ratio of particle surface area to volume, the hydraulic radius is defined as the ratio of this volume $\epsilon/(1 - \epsilon)$ to the surface σ when both values are related to the same volume of particles:

$$r_h = \frac{\epsilon}{(1 - \epsilon)\sigma} \tag{6-7}$$

For spherical particles σ is equal to the ratio of the surface area d^2 to the volume $\pi d^3/6$, i.e., $\sigma = 6/d$. Hence:

$$r_h = d\frac{\epsilon}{(1 - \epsilon)6} \tag{6-8}$$

From the analysis of Cheremisinoff et al.[1], the settling velocity of spherical particles is:

$$u = u_p \left[\epsilon^2 \times 10^{-1.82(1-\epsilon)}\right] \tag{6-9}$$

For more thickened sludges, i.e., at $\epsilon < 0.7$:

$$u = \left(0.123\frac{\epsilon^3}{(1 - \epsilon)}\right)u_p \tag{6-10}$$

The above equations apply strictly to discrete particle settling. In contrast to single particle settling, actual systems form a certain structural unity similar to tissue. The sludge is compacted under the action of gravity force (i.e., the void fraction decreases and the liquid is squeezed out from the pores). The formation of a regular sediment from a flocculent may be achieved by the addition of electrolytes.

The general characteristics of normal settling of nonspherical particles and flocculent is that the sediment carries along with it a portion of the liquid by trapping it between particle cavities. This trapped volume of liquid flows downward with the sludge and is proportional to the volume of the sludge. That is, it can be expressed as $a(1 - \epsilon)$, where "a" is a coefficient, and $(1 - \epsilon)$ is the volume of particles. Consequently, a portion of the liquid remains in a layer above the sludge, and a portion is carried along with the sludge corresponding to the modified void fraction:

$$\epsilon' = \epsilon - a(1 - \epsilon) \tag{6-11}$$

This is the difference of the total relative liquid volume and liquid moving together with particles.

The settling velocity at $\epsilon < 0.7$ is:

$$u = 0.123(1 + a)^2 u_p \frac{\left(\epsilon - \dfrac{a}{1 + a}\right)^3}{1 - \epsilon} \tag{6-12}$$

or letting $a/(1 + a) = \beta$,

$$u = \frac{0.123u_p(\epsilon - \beta)^3}{(1 - \beta)^2(1 - \epsilon)} \qquad (6\text{-}13)$$

For slurries with nonspherical particles:

$$u = u_p \frac{(\epsilon - \beta)^2}{1 - \beta} 10^{-1.82\frac{1-\epsilon}{1-\beta}} \qquad (6\text{-}14)$$

Clarification capacity is based on the settling velocity of the suspended solids. Sedimentation tests are almost always recommended when scaling up for large settler capacities. By the material balance method described earlier, the total amount of fluid is equal to the sum of the fluid in the clear overflow plus the fluid in the compacted sludge removed from the bottom of the thickener. The average vertical velocity of fluid at any height through the thickener is the volumetric rate passing upward at that level divided by the unit's cross section. If the particle settling velocity is less than the upward fluid velocity, particles will be entrained out in the overflow. For those size particles whose settling velocity approximately equals that of the upward fluid velocity, particles remain in a balanced suspension and the solid's concentration in the clarification zone increases. This results in a reduction of the settling velocity until the point where particles are entrained out in the overflow. Thickeners must be designed such that the settling velocity is significantly greater than the upward fluid velocity.

Solids concentration varies over the thickener's height. At the lower levels where the solution is dense, settling is retarded. The upward fluid velocity can exceed the particle settling velocity irrespective of whether this condition exists in the upper zone or not. Figure 6-6 illustrates this where curve II denotes a higher feed rate. Proper design is based on an evaluation of the settling rates at different concentrations as compared to the vertical velocity of the fluid.

If the feed rate exceeds the design's maximum, particulates are unable to settle out of the normal clarification zone. Hence, there is an increase in the solids concentration, resulting in hindering settling. The result is a corresponding decrease in the sedimentation rate below that observed for the feed slurry. The feed rate corresponding to the condition of just failing to initiate hindered settling represents the limiting clarification capacity of the system. That is, it is the maximum feed rate at which the suspended solids can attain the compression zone. The

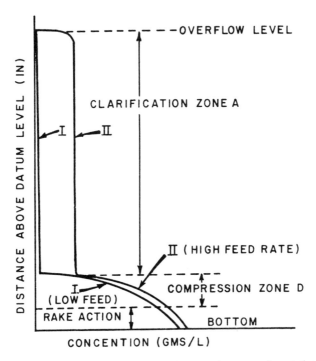

Figure 6-6. Plot of concentration vs. height in a continuous sedimentation unit. Curves are for a low feed rate (I) and a high feed rate (II).

proper cross-sectional area can be estimated for different concentrations and checked by batch sedimentation tests on slurries of increasing concentrations.

Figure 6-7 shows the effect of varying the underflow rate on the *thickening capacity*. The depth of the thickening zone (compression zone) increases as the underflow rate decreases, and hence, the underflow solids concentration increases, based on constant feed rate. The curves of concentration as a function of depth in the compression zone are essentially vertical displacements of each other and are similar to those observed for batch sedimentation. When the sludge rakes operate, they essentially break up a semirigid structure of concentrated sludge. Generally, this action extends to several inches above the rakes and contributes to a more concentrated underflow.

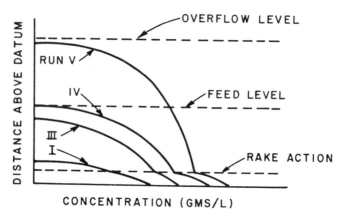

Figure 6-7. Plot showing effect of underflow rate (i.e., rate of sludge removal) on thickening capacity. The rate of sludge removal decreases from I to V.

The required compression zone height can be estimated from batch settling tests. The first batch test should be conducted with a slurry having an initial concentration equivalent to that of the top layer of the compression zone during the period of constant settling (called *critical concentration*). The time for the sample slurry to pass from the critical concentration to the desired underflow concentration is the retention time for the solids in the continuous operation.

The slope of the compression curve is:

$$\ell n \, (Z - Z_\infty) = -kt + \ell n \, (Z_c - Z_\infty) \qquad (6\text{-}15)$$

Z_c is the height of the compression zone at its critical concentration. Equation 6-15 is that of a straight line and is normally plotted as log $(Z - Z_\infty)/(Z_0 - Z_\infty)$ versus time (where Z_0 is the initial slurry concentration).

If batch tests are performed with an initial slurry concentration below that of the critical, the average concentration of the compression zone will exceed the critical value since it will consist of sludge layers compressed over varying time lengths. Here is a method for estimating the required time to pass from the critical solids content to any specified underflow concentration:

1. Extrapolate the compression curve back to the critical point or zero time.
2. Locate the time when the upper interface (between the supernatant liquid and slurry) is at height Z_o', halfway between the initial height Z_o and the extrapolated zero-time compression-zone height Z_o'.
3. This time represents the period when all the solids were at the critical dilution and went into compression. The retention time is computed as $t - t_c$, where t is the time when the solids reach the specified underflow concentration.

The procedure is illustrated in Figure 6-8. The required volume for the compression zone should be based on estimates of the time each layer has been in compression. The volume for the compression zone is the sum of the volume occupied by the solids plus the volume of the entrapped fluid. An approximate formula for this calculation is:

$$V = Q\Delta t \left[\frac{1}{\rho_s} + \frac{1}{\rho_\ell} \left(\frac{m_\ell}{m_s} \right)_{avg} \right] \qquad (6\text{-}16)$$

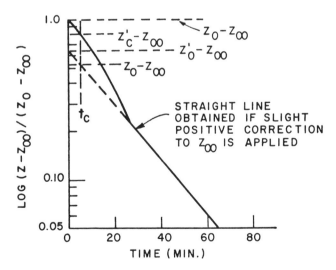

Figure 6-8. Illustrates the extrapolation of sedimentation data to estimate the time for critical concentration.

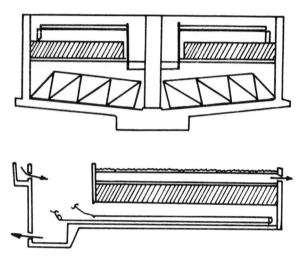

Figure 6-9. Tube settlers in existing clarifier.

Shallow Settlers

Shallow settling devices such as tube settlers have been applied to water and wastewater treatment. Tube settlers consist of bundles of small tubes with hydraulic radii ranging from one inch upward and lengths of 2 ft or more, depending upon the particular application. Square tube sections are most common but hexagonal and other shapes have been used.

Tubes are inclined steeply (60°) to the horizontal and fabricated in modules. These modules have beam strength which permits installation in settling tanks (see Figure 6-9). Clarifier influent is fed beneath the tube modules. The flow passes upward through the modules with the solids moving counter-currently by gravity and falling from the tube bottoms into the sludge collection zone beneath. The clarified effluent is collected above the tube modules.

Tube settlers promote sedimentation by:

1. The multiple tubes stacked one above another provide effective settling area several times that of the projection in plan of the modules.

2. The small hydraulic radius of the tubes maintains laminar flow and promotes uniform flow distribution.
3. In steeply inclined tubes, the movement of sludge against the direction of flow favors particle contact and agglomeration. This additional flocculation offsets the reduction in their horizontal projected area caused by inclining the tubes. Tube settlers have been promoted both for reducing required size of settling tanks and for improving performance.

Tube settlers have found wider application in water treatment than in wastewater. For wastewater, tube settlers find best applications in tertiary coagulation and settling. They also may assist in upgrading performance of units with serious short circuiting.

When installed, settling tubes usually cover one half to two thirds of the basin area. To prevent tube fouling, the remaining areas between the inlet and tube area are arranged to provide scum removal. The portion of the basin equipped with settlers should have collecting weirs at 15 ft or closer spacing to induce an even vertical flow distribution and reduce short circuiting.

The main parameter affecting settling tank performance is the surface hydraulic loading. (Q/A). This is the inflow (Q) divided by the surface area (A) of the basin (commonly expressed in gpd/sq ft).

Performance is a function of surface loading alone under conditions of quiescent or nonturbulent flow, uniform distribution of velocity over all sections normal to the general flow direction, discrete noninteracting particles, and/or no resuspension of particles once they reach the floor of the basin.

Under these conditions all particles whose settling velocity (V_s) exceeds Q/A are removed. In addition, in horizontal flow tanks particles of lower settling velocities are partially removed in the proportion $V_s/(Q/A)$.

In actual basins conditions depart from these idealized conditions because:

• Currents induced by inlets, outlets, wind and density differences may cause short circuiting or dead spaces within the tank.
• Turbulence due to forward velocity or currents in the tank retards settling.
• Flocculent solids may agglomerate into larger particles during passage through the basin.
• Sludge may be scoured and resuspended at high forward velocities.
• When influent solids concentrations are high, particles settle as a mass rather than discretely.

To account for departures of full scale tanks from ideal or test conditions, safety factors in the following ranges are recommended[3]:

Sizing Parameter	Design Factor
Area	1.25 to 1.75
Volume	1.5 to 2.0

Safety factors are not intended to cover extreme variations in flows or solids loadings, or to allow for operation at temperatures significantly different from those in the tests. Neither do they include standby capacity as needed for units critical to overall plant performance.

Short circuiting greatly reduces removal efficiencies of settling tanks. Effects are most critical for flocculant suspensions whose removal is affected by detention time. Short circuiting is accentuated by high inlet velocities, high outlet weir rates, close placement of inlets and outlets, exposure of tank surface to strong winds, uneven heating of tank contents by sunlight, and density differences between inflow and tank contents. Inlet and outlet conditions, tank geometry, and density differences due to influent SS concentrations produce steady short circuiting, whereas effects of other factors are generally intermittent.

The degree of short circuiting in circular units varies considerably depending on the type of inlet. Inlet conditions are more critical than those at the outlets. Good designs avoid short circuiting as much as possible. Major factors to consider in controlling short circuiting are dissipation of inlet velocity, protection of tanks from wind sweep and uneven heating, and reduction of density currents associated with high inlet SS concentrations.

Turbulence levels in settling basins are difficult to estimate. The only exception is turbulence due to drag from net forward velocity. Camp[4] gives a basis for estimating turbulence and for compensating for its effects by increasing tank area. Required increases vary directly with forward velocity in the tank and with the removal rate.

For clarifiers, these general guidelines should be applied:

- Provide for even inlet flow distribution to minimize inlet velocities and short circuiting.
- Minimize outlet currents and their effects by limiting weir loadings.
- Provide sufficient sludge storage depths to permit desired thickening of sludge.
- Provide sufficient wall height to give a minimum of 18 in. of freeboard.

- Reduce wind effects on open tanks by providing wind screens and by limiting fetch of wind on tank surface with baffles, weirs, or launders.
- Maintain equal flow to parallel units. Equal flow distribution between settling units is generally obtained by designing equal resistances into parallel inlet flow ports or by flow splitting in symmetrical weir chambers. The selection of *primary sedimentation* largely depends on economics.

In the absence of reliable performance-loading relations, Table 6-1 can be used as a rough guide to primary settling tank designs. Parameters listed in Table 6-1 are applicable to normal municipal wastewaters and should provide SS removals of 50 to 60%.

Weir loading limitations between 10,000 and 30,000 gpd/ft (24-hr basis) have been suggested for primary tanks[5]. At surface loadings up to 1,200 gpd/sq ft, round tanks with single peripheral weirs are used. These generally are of large diameters (> 100 ft). Normal practice is to provide a single weir. In contrast, at surface loadings as low as 600 gpd/sq ft rectangular tanks with single transverse weirs across the effluent end exceed this range if the tank length is over 50 ft. Although rectangular tanks with weir rates of more than 100,000 gpd/ft have shown SS removal in the normal range, rectangular tanks are commonly equipped with multiple weir troughs to provide loadings of 30,000 gpd/ft or less. Weir loadings are not as critical for primary tanks as they are for secondary clarifiers.

Sludge solids can be estimated directly from the expected SS removal. Sludge volume can be estimated based on expected concentration. If sludge is properly thickened in the primary tank, solids concen-

Table 6-1
Design Parameters for Primary Clarifiers

Type of Treatment	Hydraulic Loading		Depth
	Average	Peak	
	gpd/sq ft		ft
Primary Settling Followed by Secondary Treatment	800-1,200	2,000-3,000	10-12
Primary Settling with Waste Activated Sludge Return	600-800	1,200-1,500	12-15

Table 6-2
Typical Design Parameters for Secondary Clarifiers

| Type of Treatment | Hydraulic Loading | | Solids Loading | | Avg. Depth |
| | Average | Peak | Average | Peak | |
	gpd / sq ft		lb solids / day / sq ft		ft
Settling Following Trickling Filtration	400-600	1,000-1,200	—	—	11
Settling Following Air Activated Sludge (Excluding Extended Aeration)	400-800	1,000-1,200	25	50	13
Settling Following Extended Aeration	200-400	800	25	50	13
Settling Following Oxygen Activated Sludge with Primary Settling	400-800	1,000-1,200	30	50	13

trations of 2 to 7% can be achieved. On this basis typical primary sludge volumes for domestic sewage would range from 0.2 to 0.5% of plant flow. The concentration used in particular estimates should be based on actual plant experience or at least on settling/thickening tests.

Clarifiers following *trickling filters* are based on the hydraulic loading. Solids loading limits are not used in this sizing. Typical design parameters for clarifiers following trickling filters are given in Table 6-2. In applying hydraulic loading values, sizing should be calculated for both peak and average conditions and the largest value determined should be used.

Activated sludge settling tanks serve two roles: solids separation and production of a concentrated return flow to sustain biological treatment. Figure 6-10 shows the preferred arrangement.

The initial separation of activated sludge solids involves hindered rather than discrete settling. For this type of settling, tanks must be sized so the maximum surface hydraulic loading is less than the minimum initial settling velocity (ISV) expected at maximum mixed liquor concentration and at minimum temperature. If the hydraulic loading exceeds the ISV, overflow of solids will result.

To produce a concentrated return flow, activated sludge settling tanks must be designed to meet thickening as well as solids separation requirements. The rate at which solids are transported downward and removed in the tank underflow is an important consideration. This is termed the solids transport or solids flux capacity, in units of solids loading (in lb/ft²/day). When the actual solids loading applied to a tank exceeds its transport capacity, solids are being added faster than they are being removed. If this condition persists, the blanket of solids in the tank will build up and eventually overflow.

The *initial settling velocity* (ISV) at actual mixed liquor concentration can be evaluated from a plot of height of the sludge-liquid interface vs. time and noting the slope of the straight line portion of the plot. The critical minimum ISV value for a particular system may be estimated from results of a number of individual tests. One should establish relations between ISV and biological process parameters such as mixed liquor concentration and organic loading. The selected ISV value should then reflect conditions most unfavorable to settling, including correction for minimum expected temperature.

The resulting maximum surface hydraulic loading should not be exceeded by any sustained maximum flow. For good settling of activated

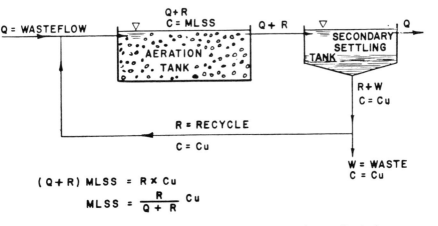

Figure 6-10. Dependence of MLSS concentration on secondary settling tank underflow concentration.

sludges from municipal wastewaters, this design relation between the ISV and the mixed concentration is suggested[6]:

$$V_i = 22.5e^{-338C_i}$$ (6-17)

where: V_i = settling velocity in ft/hr
C_i = concentration in lb/lb

Another settling parameter used in wastewater treatment is the *sludge volume index* (SVI). SVI provides an indication of sludge compaction characteristics. The index is calculated by dividing the initial mixed liquor SS concentration (percent) into the settled volume (percent of initial volume) occupied by the solids after one half hour of settling.

The reciprocal of the SVI can be taken as an indication of the maximum return sludge concentration which can be obtained with a given mixed liquor (100/SVI = percent solids). The index can be used as a guide to sizing return sludge pumping requirements to maintain different mixed liquor concentrations. SVI does not give a direct indication of solids transport capacity.

References

1. Cheremisinoff, N. P. and D. S. Azbel, *Fluid Mechanics and Unit Operations,* Ann Arbor Science Pub., Ann Arbor, MI (1983).
2. Cheremisinoff, P. N. and R. A. Young, *Pollution Engineering Practice Handbook,* Ann Arbor Science Pub., Ann Arbor, MI (1975).
3. O'Connor, D. J. and W. W. Eckenfelder, "Evaluation of Laboratory Settling Data for Process Design," in *Biological Treatment of Sewage and Industrial Wastes,* Vol. 2, Rheinhold Pub. Co. (1958).
4. Camp, T. R., *Trans. Am. Soc. Civil Engrs.,* 111, 895 (1946).
5. Fair, G. M., J. C. Geyer and D. A. Okun, *Water and Wastewater Engineering,* 2, John Wiley & Sons, NY (1968).
6. *Suggested Peaking Considerations for Activated Sludge,* Sanitary Engineering Staff Report, Iowa State University (1971).
7. Dick, R. I. and K. W. Young *Proceedings of 27th Purdue Industrial Waste Conference* (1972).

7

CENTRIFUGES

Sedimentation Centrifuges

In centrifuges, liquid and solids are acted upon by two forces; gravity acting downward and centrifugal force acting horizontally. In commercial units, the centrifugal force component is normally so large that the gravitational component may be neglected.

Tables 7-1 through 7-3 contain general information useful in selecting centrifuge types. Table 7-1 lists typical operating ranges. Table 7-2 gives comparative performances of different centrifuges. Table 7-3 provides a relative rating of centrifuge types in terms of suitability for sludge dewatering operations.

The magnitude of the centrifugal force component is defined by the ratio $R_c/R_g = \omega^2 r/g$ ("Relative Centrifugal Force" (RCF) or *Centrifugal Number* (N_c) (where $R_c = m\omega^2/2$ or $R_c = m\dot{\nu}^2/2$ is the centrifugal force, and $R_g = mg$ is the gravitational force). The RCF typically varies from 200 times gravity for large basket centrifuges to 360,000 for ultracentrifuges.

For liquid-solid separations, centrifugal force may be applied in either sedimentation-type centrifuges, in centrifugal filters, or a combination of both.

Principles of centrifugation are illustrated in Figure 7-1. In Figure 7-1A a stationary cylindrical bowl contains a suspension of solid particles where $\rho_\rho > > \rho_\rho$. Because the bowl is stationary the free liquid surface is horizontal and the particles settle due to gravity. In Figure 7-1B, the bowl rotates about its vertical axis. Liquid and solid particles are acted upon by gravity and centrifugal forces, resulting in the liquid assuming a position with an almost vertical inner surface (free interface).

Table 7-1
Centrifuge Types and Typical Operating Ranges

Type Centrifuge	Bowl dia. Range (in.)	Centrifugal Number	Typical Capacity Range	Method of Solids Discharge
Filtering Type				
Conical Screen—Wide angle	—	≤ 1,400	15,000 gph	C
—Differential scroll	—	≤ 1,800	70 tons/hr solids	C
—Vibrating screen	—	≤ 500	100 tons/hr solids	C
—Pusher	—	1,900	10 tons/hr solids	B
Cylindrical Screen Type				
Pusher	—	1,500	40 tons/hr solids	C
Differential Scroll	—	1,500	40 tons/hr solids	C
Horizontal	—	1,300	25 tons/hr solids	B
Vertical	—	900	10 tons/hr solids	B
Sedimentation Type				
Tubular	2–6	60,000	3,000 gph	B
Disc	9–35	2,500–8,000	12,000–24,000 gph	B–SC
Solid Bowl (constant speed horizontal)	14–38	1,000–3,500	60 ft³/batch	B
Solid Bowl (variable speed vertical)	12–90	≤ 3,200	15 ft³/batch	B
Solid Bowl (continuous)		≤ 3,200	≤ 65 tons/hr solids	C

C—Continuous
SC—Semi-continuous
B—Batch

Table 7-2
Comparative Performances of Different Centrifuges
(Data of Ambler[3])

		Σ_c Values (ft^2)	
	Calculated from Geometry	From Experimental Data Clarifying Ideal Systems	Extrapolation on Commercial Systems (Supercentrifuge Tests)
Laboratory supercentrifuge (tubular bowl 1¾ in. I.D. x 7¼ in. long) operating at:			
10,000 rpm	582	582	582*
16,000 rpm	1,485	1,485	
23,000 rpm	3,070	3,070	1,290
50,000 rpm	14,520	14,520	not used
No. 16 supercentrifuge (tubular bowl 4⅛ in. I.D. x 29 in. long) operating at:			
15,000 rpm	27,150	27,150	27,150
No. 2 disk centrifuge, 1⅞ in. r_1 x 5¾ in. r_2 on disks			
52 disks, 35° half angle, 6000 rpm	178,800	98,000	89,400 to 178,800
50 disks, 45° half angle	134,000	72,600	67,900 to 134,000
Super-D-Cantor (solid-bowl centrifugal) PN-14 (conical bowl) 3250 rpm			
(D = 14 in.–8 in., L = 23 in.)	4,750	2,950	2,950*
PY-14 (cylindrical bowl), 3250 rpm			
(D = 14 in., L = 23 in.)	8,940	5,980	5,980*

* For relatively low throughput rates.

Σ_c = capacity factor (product of area of cylindrical surface of sedimentation in the rotor and the centrifugal number)

N_c = centrifugal number = $\dfrac{u^2}{gr}$

Table 7-3
Suitability of Centrifuges to Sludge Dewatering Applications

Type Sludge	Basket Centrifuge	Conveyor Centrifuge (Solid Bowl)	Disc Centrifuge with Nozzle Discharge
• Sewage, primary raw	G	G	NA
• Sewage, primary aerobic digested		F	NA
• Sewage, secondary biological			
(activated & humus)	G		G
(activated—with alum)	G	F	G
• Sewage, primary and secondary			
biological, cosettled	G	F	NA
• Sewage, whole, aerobic digested	G	F	
• Sewage, primary, heat-treated raw	G	G	NA
• Industrial, coarse solids	F	G	NA
• Industrial, clean biological	G		G
• Industrial, hydrous or			
flocculant solids	G	F	F
• Industrial, oil-water emulsion	F	NA	G
• Industrial, fine solids	G	F	G
• Water treatment with alum	G	F	G

If the suspension consists of several components, each with different densities, they will stratify with the lightest component nearest the axis of rotation and the heaviest adjacent to the solid bowl wall (Figure 7-1C).

In Figure 7-1D, the bowl wall is perforated and lined with a permeable membrane, such as filter cloth or wire screen which will support and retain the solid particles but allow the liquid to pass through.

The main components are:

1. A rotor or bowl in which the centrifugal force is applied to a heterogeneous system to be separated.
2. A means for feeding this system into the rotor.
3. A drive shaft.
4. Axial and thrust bearings.
5. A drive mechanism to rotate the shaft and bowl.
6. A casing or "covers" to contain the separated components.
7. A frame for support and alignment.

The *tubular-bowl centrifuge* has a centrifugal number in the range of 13,000 and is designed for low capacities (50–500 gal/hr). It can handle only small concentrations of solids.

The *solid-bowl centrifuge* has maximum bowl diameters ranging from 4 to 54 in. Larger-diameter machines can handle up to 50 tons/hr of solids with a centrifugal number up to 3,000. Similar centrifuges are manufactured with a perforated wall on the bowl. These machines operate like filters with the filtrate draining through the cake and bowl wall into a collector.

The *disk-bowl centrifuge* is larger than the tubular-bowl centrifuge and rotates at slower speeds with a centrifugal number up to 14,000. These machines are capable of handling as much as 30,000 gal/hr of feed containing moderate quantities of solid particles.

Tubular-bowl centrifuges are extensively used for the purification of oils by separating suspended solids and free moisture from them; for removal of oversize particles from dye pastes, pigmented lacquers, and enamels; for "polishing" citrus and other aromatic oils, and other small-scale separating applications.

The tubular bowl-type clarifiers and separators are comprised of small-diameter cylinders (about 100 mm). Commercial machines typically work at 15,000–19,000 rpm (N_c = 13,000–18,000). Figure 7-2 shows the operating principles for these type centrifuges. See Cheremisinoff et al.[1] for explanation of operating principles.

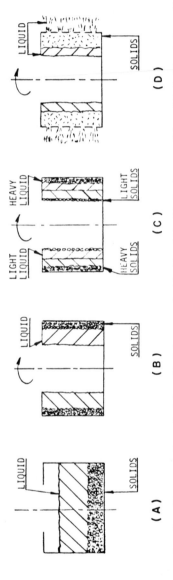

Figure 7-1. Centrifuge operating principles: (A) stationary cylindrical bowl (liquid plus heavy solid particles); (B) rotating cylindrical bowl (liquid plus heavy solid particles); (C) Rotating cylindrical bowl (two liquids plus light and heavy solid particles); (D) rotating perforated bowl.

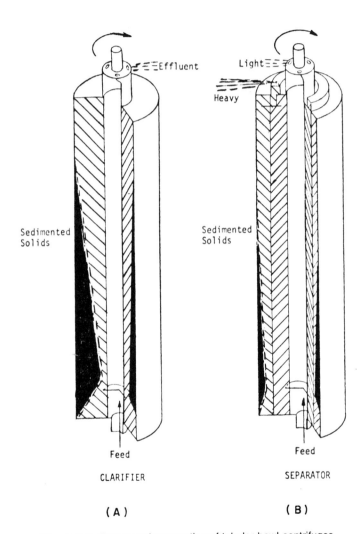

Figure 7-2. Illustrates the operation of tubular bowl centrifuges.

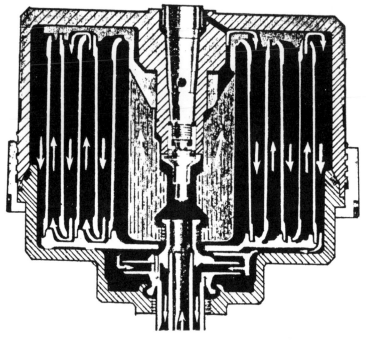

Figure 7-3. Illustrates a Multipass Centrifuge.

Multichamber (Multipass) Centrifuges combine the process principles of a tubular clarifier with mechanical drive, and the bowl contour
of a disk centrifuge. The suspension flows through a series of nested
cyinders of progressively increasing diameter. The direction of the flow
from the smallest to the largest cylinders is in parallel to the axis of
rotation as in the tubular bowl (see Figure 7-3).

The rotor usually contains six annuli, so that the effective length of
suspension travel is approximately six times the interior height of the
bowl. The design can be considered as a multistage classifier because
the centrifugal force which acts in the machine is greater in each subsequent annulus. Consequently, larger, heavier particles are deposited in
the first annulus, and lighter particles are deposited in the last annulus
(the zone of greatest centrifugal force). The radial distance particles
must migrate to reach the cylinder wall is thus minimized. Multipass
rotors typically have a total holding volume of up to 65 liters for the

largest size, of which about 50% is available for the retention of collected solids before the clarification process is impaired. These machines are applied to the clarification of fruit and vegetable juices, wine, and beer.

The continuous solid-bowl centrifuge consists of a solid-wall rotor which may be tubular or conical in shape, or a combination of the two. The rotor may rotate about a horizontal or a vertical axis because the centrifugal force is many times that of gravitational force (in many units $N_c > 3000$).

A typical example of this kind of equipment is a continuous horizontal centrifuge as shown in Figure 7-4. The cylindrical rotor has a truncated cone-shaped end and an internal screw conveyor rotating together. The screw conveyor often rotates at a rate of one or two rpm below the rate of rotation of the rotor. The suspension enters the bowl axially through the feed tube to a feed accelerated zone, where it passes through a feed port in the conveyor hub into the pond. The suspension is subjected to centrifugal force and thrown against the bowl wall where the solids are separated. The clarified suspension moves towards the broad part of the bowl to be discharged through a port.

In tubular- and solid-bowl centrifuges, to evaluate the radial velocity of a particle moving toward a centrifuge wall, the expression for particle settling in a gravitational field is applied with the acceleration term replaced by centrifugal acceleration "a":

$$a = \frac{u_r^2}{r} \tag{7-1}$$

where: u_r = the peripheral velocity at a distance r from the axis of rotation.

In terms of the number of rotations n:

$$u_r = 2\pi r n \tag{7-2}$$

Centrifugal acceleration is:

$$a = 4\pi^2 r n^2 \tag{7-3}$$

The motion of large particles at Re > 500 is turbulent; the settling velocity is:

$$u_r = 1.74 \sqrt{\frac{d(\gamma_p - \gamma)a}{\gamma}} \tag{7-4}$$

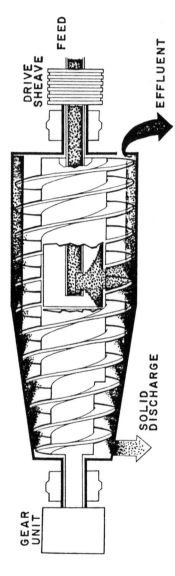

Figure 7-4. Illustrates a continuous solid-bowl centrifuge.

where: γ_p = solid's specific weight
γ = liquid's specific weight

The particle velocity in the radial direction of the wall is:

$$u_r = 10.94 \sqrt{\frac{d(\gamma_p - \gamma)r}{\gamma}} \qquad (7\text{-}5)$$

The number of times the particle velocity in a centrifuge exceeds that in free particle settling (i.e., the separation factor K_s) is:

$$K_s = \frac{u_r}{u_g} = 2\pi n \sqrt{\frac{r}{g}} \qquad (7\text{-}6)$$

For example, at n = 1,200 rpm = 20 1/sec and r = 0.5 m, the settling velocity in the centrifuge is almost 28 times greater than that in free settling. Note that Equation 7-5 is only applicable at Re > 500.

For small particles (Re < 2) particulate migration towards the wall is laminar. The proper settling velocity expression is:

$$u_r = \frac{d^2(\rho_p - \rho)a}{18\mu} \qquad (7\text{-}7)$$

or

$$u_r = \frac{2\pi^2 d^2(\gamma_p - \gamma)n^2 r}{9\mu g} \qquad (7\text{-}8)$$

The separation factor for this regime is:

$$K_s = \frac{u_r}{u_g} = 4\pi^2 n^2 \frac{r}{g} \qquad (7\text{-}9)$$

For the same case of n = 1,200 rpm and r = 0.5, u_r/u_g = 800, the turbulent regime ratio was only 28. This example demonstrates that the centrifugal process is more effective in the separation of small particles rather than large ones. Note that after the radial velocity u_r is determined, it is necessary to check whether the laminar condition (Re < 2) is fulfilled.

For the transition regime (2 < Re < 500) the expression for the particle radial velocity towards the wall is:

$$u_r = \frac{0.153 d^{1.14}(\rho_p - \rho)^{0.71} a^{0.71}}{\rho^{0.29} \mu^{0.43}} \qquad (7\text{-}10)$$

or

$$u_r = \frac{1.36 d^{1.14}(\gamma_p - \gamma)^{0.71}(n^2 r)^{0.71}}{\gamma^{0.29}(\mu g)^{0.43}} \qquad (7\text{-}11)$$

The ratio of settling velocity in a centrifuge to that in the gravitational field is:

$$K_s = \frac{u_r}{u} = \left(4\pi^2 n^2 \frac{r}{g}\right)^{0.71} \qquad (7\text{-}12)$$

This ratio represents an average between similar ratios for the laminar and turbulent regimes.

For the most general case, $u_r = f(D, \rho_p, \rho, \mu, \omega, r)$, where we ignore the flow regime, the dimensionless centrifugal Archimedes number is defined as:

$$Ar_c = C_D Re^2 = \frac{4}{3} \frac{d_p^3 \rho (\rho_p - \rho)\omega^2 r}{\mu^2} \qquad (7\text{-}13)$$

From Ar_c the plot given in Figure 7-5 may be used to evaluate the Reynolds number, and subsequently, u_r. The following example from Loncin[2] illustrates the use of Figure 7-5.

Sample Calculation 7-1.

Oil droplets ($d_p = 10^{-4}$ m, $\rho_p = 900$ kg/m^2) suspended in water ($\rho = 1,000$ kg/m^3, $\mu = 10^{-3}$ decipoise) are separated in a sedimentation centrifuge designed to operate at 5,000 rpm ($\omega = 2\pi \times 5,000/60$). If the distance of a single droplet from the axis of rotation is 0.1 m, determine the droplet's radial settling velocity.

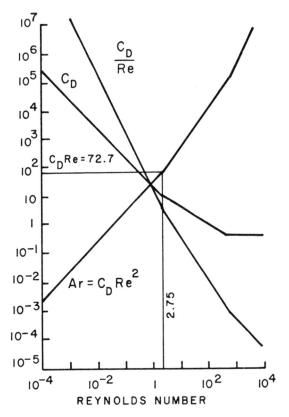

Figure 7-5. Plot of Archimedes number versus Reynolds number.

Solution:

$$Ar_c = C_D \times Re^2$$

$$= \frac{4 \times 10^{-12} \times 1{,}000 \times 100 \left(\dfrac{2\pi \times 5{,}000}{60}\right)^2}{3 \times 10^{-6}} \times 0.1$$

$$= 3{,}650$$

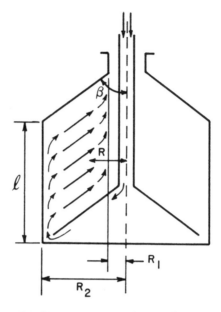

Figure 7-6. Shows schematic of a centrifuge operation.

An absolute value of $(\rho_p - \rho)$ is assumed. A negative value indicates that the droplet displacement is centripetal. The value of the Reynolds number corresponding to $Ar_c = 3{,}650$, from Figure 7-5, is 45. The radial settling velocity is evaluated from Re:

$$45 = \frac{10^{-4} \times 1{,}000 \; u_r}{10^{-3}}$$

or

$$u_r = 0.45 \text{ m/sec.}$$

If the separation were to occur in a gravitational field only, the droplet velocity would be $u = 5.45 \times 10^{-4}$ m/sec. For laminar droplet motion this corresponds to a separation number of 2,800. However, in this case, the flow is in the transition regime and $K_s = 825$.

To estimate the capacities of tubular- and solid-bowl centrifuges, consider the following. When a rotating centrifuge is filled with suspension, the internal surface of liquid acquires a cylindrical geometry of radius R_1 (see Figure 7-6). If the liquid is lighter than the solid particles, the liquid moves towards the axis of rotation while the solids flow towards the bowl walls. A simplified model of centrifuge operation is that of a cylinder of fluid rotating about its axis. The flow forms a layer bound outwardly by a cylinder, R_2, and inwardly by a free cylindrical surface, R_1. This surface is, at any point, normal to a resulting force (centrifugal and gravity) acting on the solid particle in the liquid. The gravity force is in general negligible compared to the centrifugal force and the surface of liquid is perpendicular to the direction of centrifugal force.

For a solid particle located at distance R from the axis of rotation, the particle moves centrifugally with a settling velocity u_s while liquid particles move in the opposite direction centripetally with a velocity u_f:

$$u_f = \frac{\dfrac{dV}{d\tau}}{2\pi R \ell} \tag{7-14}$$

where V = volume (m³;) τ = time (sec); ℓ = height of a bowl. The resulting velocity will be centrifugal and the solid particles will be separated when:

$$u_s > \frac{\dfrac{dV}{d\tau}}{2\pi R \ell} \tag{7-15}$$

The capacity is:

$$\frac{dV}{d\tau} = 2\pi R \ell u_s \tag{7-16}$$

If the particle's density is lower than that of the liquid the path of the liquid is centrifugal. Settling will occur when u_s (centripetal) is higher than the radial velocity u_f (centrifugal).

Settling capacity for a given size of particle is a function of ℓ, R, and u_s, which is itself proportional to R. In general, for the sedimentation of heavy particles in a suspension, it is sufficient that the radial component of u_f be less than u_s at a radius greater than R_2.

Because of turbulence effects it is generally good practice to limit the settling capacity so that u_s again exceeds u_f near R_1. The same situation occurs when the particles are lighter than the continuous liquid.

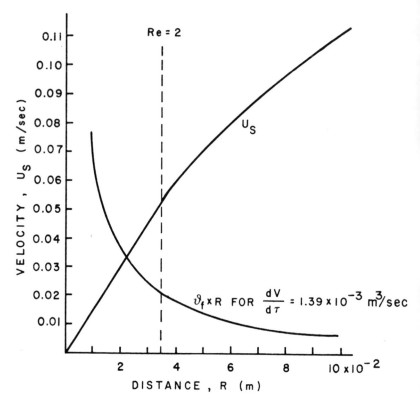

Figure 7-7. Plots of u_s and $u_f R$ vs. R.

Sample Calculation 7-2.

Estimate the settling velocity of a particle ($d_p = 4 \times 10^{-5}$ m and density $\rho_p = 900$ kg/m^3) through water in a sedimentation centrifuge operating at a velocity $\omega = 4{,}000$ rpm. The particle velocity is a function of its distance from the axis of rotation. Data are provided in Figure 7-7.

Solution. The data indicate that the particle motion is laminar for values of R up to 0.34 m. For R values exceeding 0.04, u_s must be estimated from the Reynolds-Archimedes relationship. As part of the analysis let's consider similar spherical particles of diameter 20×10^{-5} m being centrifuged at the same conditions at a rotating velocity of

$\omega = 523$ radian per minute (5,000 rpm). The settling velocities for such a particle are plotted in Figure 7-8. The particles up to a distance of 0.44 m settle in the transition regime. At higher values of R the regime is turbulent.

To evaluate the capacity, or more correctly, the behavior of a fixed capacity $dV/d\tau$:

$$u_f R = \frac{\frac{dV}{d\tau}}{2\pi\ell}$$

If, for example $dV/d\tau = 1.39 \times 10^{-3}$ m³/sec (5,000 liters per hour) and ℓ (height of the bowl) = 0.3 m, we have:

$$u_f R = \frac{1.39 \times 10^{-3}}{2\pi \times 0.3} = 0.736 \times 10^{-3}\, m^2/sec$$

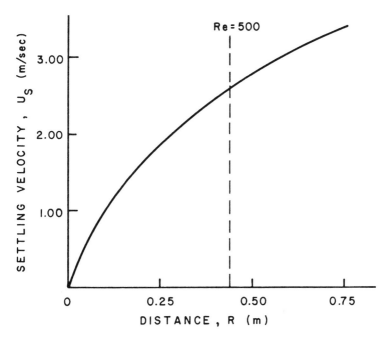

Figure 7-8. Settling velocity as a function of R for 200μm particles.

From Figure 7-7, for radii > 0.002 m, $u_s > u_f$ and consequently the particles of diameter 4×10^{-5} m(40μm), and density $\rho_p = 900$ kg/m³ are centrifuged at 4,000 rpm in water and are displaced centripetally.

For 200 μm particles at 5,000 rpm, the total sedimentation occurs at the same conditions for $R < 0.01$ m. The capacity of a sedimentation centrifuge can be determined from Equation 7-16 where R_1 = radius corresponding to the internal surface of the liquid; u_s = settling velocity at R_1.

Power Requirements and Applications Data

Overall power requirements for a centrifuge are based on five contributions: power needed to start the rotor, power to set the material in the bowl in motion, power required to overcome friction, power required to transfer the sediment to discharge ports or nozzles, and the power required to discharge products from the bowl.

Power required to start the rotor. If the centrifuge bowl is started from rest to some angular velocity ω over starting time τ_s, the energy spent is:

$$E_0 = \frac{1}{2} J\omega^2 \tag{7-17}$$

where J is the inertia moment.

If the cylindrical wall of the bowl is considered (that is, neglecting the other parts of the bowl), as a first approximation:

$$E_0 = \frac{\pi \ell \rho_b}{4}(R_3^4 - R_2^4)\omega^2 \tag{7-18}$$

The required horsepower for start-up is:

$$N_{s.b.} = 10^{-3} \frac{\pi \ell \rho_b (R_3^4 - R_2^4)\omega^2}{\tau_s} \tag{7-19}$$

Example 7-2:

A steel bowl ($\rho_b = 10,600$ kg/m³) has the following characteristics:

$$R_3 = 0.32 \text{ m}$$
$$R_2 = 0.30 \text{ m}$$
$$\ell = 0.30 \text{ m}$$
$$\omega = 523 \text{ radians per second (or 5,000 rpm)}.$$

Determine the required horsepower to set the centrifuge rotor in motion.

Solution

The energy spent is:

$$E = \frac{\pi \times 0.3 \times 10,600}{4}(0.32^4 - 0.30^4) \times 523^2 = 1.63 \times 10^6$$

If the required start-up time is 900 sec, then the horsepower is:

$$N = \frac{1.63 \times 10^6}{900} = 1,810 \text{ watts}$$

Starting up the material in the bowl. If the density of the material is ρ_m, then the mass of an elemental ring at distance R from the axis of rotation is $2\pi R dR H \rho_m$. Note that $(R_b^4 - R_m^4) = (R_b^2 + R_m^2)(R_b^2 - R_m^2)$ and $(R_b^2 - R_m^2)H\rho_m = M_m$, where M_m = mass of the material.

$$n_s = \frac{1}{4} \times 10^{-3}\omega^2 \frac{M_m(R_b^2 + R_m^2)}{\tau_s}; \text{ kW} \tag{7-20}$$

Power requirements imposed by friction. Bearings, belts, gears, and seals transform small amounts of power into heat. The greatest loss of power is derived from the friction of the gas in contact with the surface of the rotating elements of the centrifuge (referred to as "windage"). This value is determined as follows:

$$N_f = \dot{F}_f w \tag{7-21}$$

where:

$$\dot{F}_f = P\lambda_f, \text{ the friction force.}$$

and P = force pressing the body to the surface,
     λ_f = friction coefficient,
     w = the linear velocity of the body relative to the surface considered.

The power required to overcome friction is:

$$N_f = \frac{1}{2} \times 10^{-3} P\lambda_f \omega D \tag{7-22}$$

Individual contributions to friction are:

Friction in bearings. The horsepower expended in overcoming the friction in bearings is:

$$N_{f.b.} = \frac{1}{2} \times 10^{-3}(M_b + M_m)g\omega D_b\lambda_b, \text{ kW} \tag{7-23}$$

where D_b = diameter of the shaft neck
       λ_b = friction coefficient in the shaft journal.

For ball bearings, λ_b is typically 0.02 to 0.03.

Friction in the glands. In a horizontal, hermetically closed centrifuge where the shaft is clamped in seals:

$$N_{f.g.} = \frac{1}{2} \times 10^{-3}(M_b + M_m)g\omega D_s\lambda_s \tag{7-24}$$

where D_s and λ_s are shaft diameter and friction, respectively.

The friction coefficient in the gland assembly typically ranges from 0.4 to 0.6.

Friction of material against the conical bowl of a centrifuge. M_o = the amount of sediment (in kg) transported along the generatrix L of the bowl in a unit time (sec). The material is loaded at approximately the middle of the bowl, and under the action of centrifugal forces, is uniformly distributed over the entire surface of the bowl. R_{av} = average bowl radius.

$$N_{f.b.} = \frac{1}{2} \times 10^{-3} M_o \omega^2 R_{av} H \lambda_b; \text{ kW} \tag{7-25}$$

Friction against the screw conveyor. The sediment is transported by a screw conveyor. Let α be the angle between the fillet of the screw conveyor and the vertical axis and Z' be the number of fillets. The length of one fillet of the screw conveyor is $2\pi R_{av}/\cos \alpha$, and the total length of the fillet is $2\pi R_{av}Z/\cos \alpha$, where R_{av} is an average radius.

$$N_{f.s.} = \frac{1}{2} \times 10^{-3} \pi \omega^2 R_{av}{}^2 M_o Z' (\sin 2\alpha + 2 \cos^2 \alpha \lambda_b) \lambda_s \tag{7-26}$$

where $N_{f.s.}$ is in units of kW; λ_s = the friction coefficient between screw and sediment.

Friction against air (windage). To evaluate the power lost in windage, dimensional analysis is applied:

$$N_{f.w.} = \psi \omega^3 D_{av} H; \text{ kW} \tag{7-27}$$

where D_{av} = average diameter of a bowl
ρ = density of the medium
ω = linear velocity of a bowl
H = height of the bowl
ψ = constant = 1.15×10^{-6}

The power lost in windage can exceed that required to accelerate the feed, especially in a high pressure operation.

Power requirements for sediment transportation to the discharge ports using a screw conveyor. The power for the transportation of sediment M_o(kg/sec) along half a length of generatrix ($1/2$L) is:

$$N_o = \frac{1}{2} M_o \omega^2 R_{av}{}^2 L; \text{ watt} \tag{7-28}$$

Power requirements for the discharge of products from the bowl. The products of separation—sediment and centrifuged effluent—entrain a certain amount of energy upon leaving the bowl. The power requirements for discharge of products is:

$$N_d = \frac{1}{2} \times 10^{-3}\omega^2(M_o R_o^2 M_e R_e^2); \text{ kW} \tag{7-29}$$

where M_o = mass flow of sediment, kg/sec
          M_e = mass flow of effluent, kg/sec
R_o and R_e = radii of rotation of sediment before and after leaving the bowl, respectively.

The total power requirements for a sedimentation centrifuge is the sum of the individual power contributions:

$$N = N_d + N_o + N_{f.b.} + N_{f.s.} + N_{f.g} + N_{f.w} \tag{7-30}$$

Additional minor energy requirements include those for initial mixing of the suspension, for overcoming hydraulic resistance, for ventilation, and so on. For practical calculations, therefore, the computed total power requirements should be slightly increased.

References

1. Cheremisinoff, N. P. and D. S. Azbel, *Fluid Mechanics and Unit Operations,* Ann Arbor Science Pub., Ann Arbor, MI (1983).
2. Loncin, M., *Operations Unitaires du Genie Chimique,* Dunod, Paris (1961).
3. Ambler, C. M., *Chem. Eng. Prog.,* 48:150 (1952).

8

CHEMICAL PRETREATMENT

General Considerations

This chapter provides general information and data on chemicals commonly used for suspended solids removal. Polymers comprised of high molecular weight cationic (postively charged) polyelectrolytes are often used. Their primary role is charge reversal of either the contaminant particle or the media granual, depending on application. When polymers are used, they are generally added in the range of 1–2 ppm. This low dosage is possible because in most cases the formation of flocculants is not required; instead, the polymer influences the electrostatic charges. Excessive polymer can foul and/or mask the adhesion sites of the media. A typical plot of filtration efficiency is shown in Figure 8-1.

Polymer effectiveness depends on pH, temperature, other ions present, etc. Cationic polymers generally work best in neutral to slightly acidic solutions. Note that when a pH adjustment is needed prior to final discharge, if the stream is acidic, it should be neutralized downstream of the filter. If, however, it is caustic, it should be neutralized upstream of the filter. This same scheme of pH adjustment also has beneficial effects on the removal of emulsified oils.

Optimum polymer dosage is established at start-up. The amount of polymer addition can be varied with flow rate by use of a flow element and pneumatic stroke adjustment for the chemical feed pump. Chemical dosage is also controlled by turbidity measurements or other effluent monitoring devices (zeta potential, free oil concentration, etc.). This eliminates the necessity of overdosing the polymer during times of low flow or when contaminant levels are low.

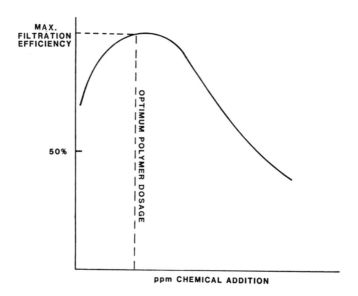

Figure 8-1. Filtration efficiency curve for polymer addition.

Properties of Common Chemicals

Properties and characteristics of common chemicals used in waste-water treatment are briefly described:

Aluminum compounds are commercially available as dry and liquid alum. Dry alum most often used in wastewater treatment is called "filter alum." $(Al_2(SO_4)_3)$; molecular weight $\simeq 600$. Alum is white to cream in color, and a 1% solution has a pH of about 3.5. The commercially available grades of alum and their corresponding bulk densities and angles of repose are:

Grade	Angle of Repose°	Bulk Density (lb/ft³)
Lump	-------	62 to 68
Ground	43	60 to 71
Rice	38	57 to 71
Powdered	65	38 to 45

Each of these grades has a minimum aluminum content of 27%, (Al_2O_3), and maximum Fe_2O_3 and soluble contents of 0.75 and 0.5%, respectively.

The solubility of commercial dry alum at various temperatures is:

Temperature (°F)	Solubility (lb/gal)
32	6.03
50	6.56
68	7.28
86	8.45
104	10.16

Dry alum is not corrosive unless it absorbs moisture from the air.

Liquid alum is shipped at a solution strength of about 8.3% as Al_2O_3 or about 49% as $Al_2(SO_4)_3 \cdot 14H_2O$.

Crystallization temperatures of various solution strengths are given in Figure 8-2.

The viscosity of various alum solutions is given in Figure 8-3.

Reactions between alum and various wastewater constituents are influenced by many factors. It is not possible to predict accurately the amount of alum that will react with a given amount of alkalinity, lime, or soda ash which may have been added to the wastewater. The reaction of Al^{3+} with OH^- ions is made available by the ionization of water or by the alkalinity of the water.

Solution of alum in water:

$$Al_2(SO_4)_3 \rightleftharpoons 2Al^{3+} + 3(SO_4)^2 \qquad (8\text{-}1)$$

Hydroxyl ions become available from ionization of water:

$$H_2O \rightleftharpoons H^+ + OH^- \qquad (8\text{-}2)$$

The aluminum ions (Al^{3+}) then react:

$$2Al^{3+} + 6OH^- \rightleftharpoons 2Al(OH)_3 \qquad (8\text{-}3)$$

Consumption of hydroxyl ions results in a decrease in alkalinity. Where the alkalinity of the wastewater is inadequate for the alum dosage, the

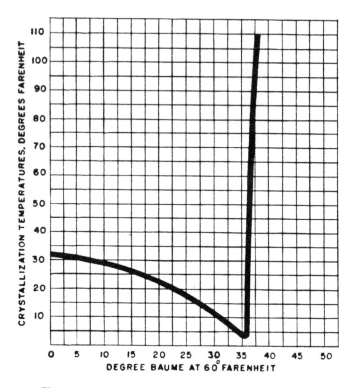

Figure 8-2. Crystallization temperatures of alum solutions.

pH must be increased by the addition of hydrated lime, soda ash or caustic soda. The reactions of alum (aluminum sulfate) with the common alkaline reagents are:

$$Al_2(SO_4)_3 + 3Ca(HCO_3)_2 \rightarrow 2Al(OH)_3\downarrow + 3CaSO_4 + 6CO_2\uparrow \quad (8\text{-}4)$$

$$Al_2(SO_4)_3 + 3Na_2CO_3 + 3H_2O \rightarrow 2Al(OH)_3\downarrow + 3CO_2\uparrow \qquad (8\text{-}5)$$

$$Al_2(SO_4)_3 + 3Ca(OH)_2 \rightarrow 2Al(OH)_3\downarrow + 3CaSO_4 \qquad\qquad (8\text{-}6)$$

For each mg/l of alum dosage, the sulfate (SO_4) content of the water will be increased approximately 0.49 mg/l and the CO_2 content of the water will be increased approximately 0.44 mg/l.

Iron compounds have pH coagulation ranges and floc characteristics similar to aluminum sulfate. They are generally corrosive and often present difficulties in dissolving. Usage may result in high soluble iron concentrations in process effluents.

Liquid ferric chloride is corrosive, dark brown, and oily in appearance {density ~ 11.2 to 12.4 lb/gal (35 to 45% $FeCl_3$)}. The pH of a 1% solution is 2.0.

The molecular weight of ferric chloride is 162.22. Viscosities of ferric chloride solutions at various temperatures are given in Figure 8-4 and freezing point curves are given in Figure 8-5.

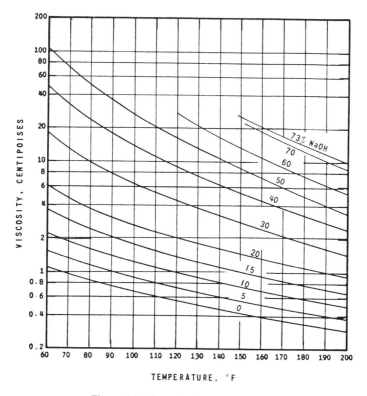

Figure 8-3. Viscosity of alum solutions.

Ferric chloride solutions are corrosive to many common materials and cause stains which are difficult to remove. Areas which are subject to staining should be protected with resistant paint or rubber mats.

Ferrous chloride ($FeCl_2$) as liquid is available as waste pickle liquor from steel processing. The liquor weighs between 9.9 and 10.4 lb/gal and contains 20 to 25% $FeCl_2$ or about 10% available Fe^{2+}. The molecular weight is 126.76. Free acid in waste pickle liquor can vary from 1 to 10% and usually averages about 1.5 to 2.0%. Ferrous chloride is slightly less corrosive than ferric chloride.

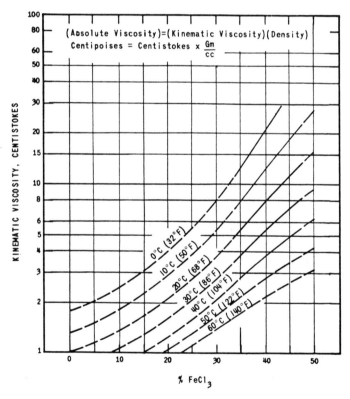

Figure 8-4. Viscosity vs. composition of ferric chloride solutions at various temperatures.

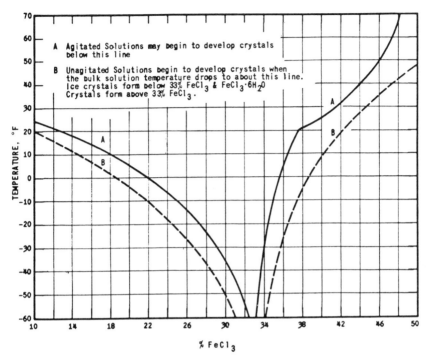

Figure 8-5. Freezing point curves for commercial ferric chloride solutions.

Ferric sulfate is available as dry, partially-hydrated granules with the formula $Fe_2(SO_4)_3 \cdot XH_2O$, where X is approximately 7. Typical properties of one commercial product are:

Molecular weight	526
Bulk density	56–60 lb/ft^3
Water soluble Fe^{+3}	19.5%
Water soluble Fe^{+2}	2.0%
Insolubles total	4.0%
Free acid	2.5%

Ferric sulfate and ferric chloride react with wastewater alkalinity or with lime or soda ash, resulting in precipitation of ferric hydroxide. Reactions using ferric sulfate are:

$$Fe_2(SO_4)_3 + 3Ca(HCO_3)_2 \rightarrow 2Fe(OH)_3\downarrow + 3CaSO_4 + 6CO_2\uparrow \quad (8\text{-}7)$$

$$Fe_2(SO_4)_3 + 3Na_2CO_3 + 3H_2O \rightarrow$$
$$2Fe(OH)_3\downarrow + 3Na_2SO_4 + 3CO_2\uparrow \quad (8\text{-}8)$$

$$Fe_2(SO_4)_3 + 3Ca(OH)_2 \rightarrow 2Fe(OH)_3\downarrow + 3CaSO_4 \quad\quad (8\text{-}9)$$

Ferric chloride can be substituted in these reactions.

Ferrous sulfate and ferrous chloride react with wastewater alkalinity or with lime to precipitate ferrous hydroxide. The ferrous hydroxide is oxidized to ferric hydroxide by dissolved oxygen in wastewater. Typical reactions are:

$$FeSO_4 + Ca(HCO_3)_2 \rightarrow Fe(OH)_2\downarrow + CaSO_4 + 2CO_2\uparrow \quad (8\text{-}10)$$

$$FeSO_4 + Ca(OH)_2 \rightarrow Fe(OH)_2\downarrow + CaSO_4 \quad\quad (8\text{-}11)$$

$$4Fe(OH)_2 + O_2 + 2H_2 \rightarrow 4Fe(OH)_3\downarrow \quad\quad (8\text{-}12)$$

Ferrous hydroxide is soluble, and oxidation to the more insoluble ferric hydroxide is necessary if high iron residuals in effluents are to be avoided. Flocculation with ferrous iron improves by addition of lime or caustic soda.

Lime refers to a variety of chemicals which are alkaline in nature and contain principally calcium, oxygen, or magnesium. Included are quicklime, dolomitic lime, hydrated lime, dolomitic hydrated lime, limestone, and dolomite.

Quicklime, CaO, has a density range of 55 to 75 lb/ft^3, (molecular weight of 56.08). A feed slurry can be prepared with up to 45% solids. Lime is only slightly soluble, and both lime dust and slurries are caustic in nature. A saturated solution of lime has a pH of about 12.4.

Hydrated lime [$Ca(OH)_2$] is a white powder (200 to 400 mesh); bulk density 20 to 50 lb/ft^3; contains 82 to 98% $Ca(OH)_2$; slightly hydroscopic. Molecular weight is 74.08. The dust and slurry of hydrated lime are caustic in nature. The pH of a saturated, hydrated lime solution is the same as that given for quicklime.

Lime differs from the hydrolyzing coagulants. When added to wastewater it increases pH and reacts with the carbonate alkalinity to precipitate calcium carbonate. If sufficient lime is added to reach a high pH, (pH ~ 10.5), magnesium hydroxide also precipitates. This latter precipitation enhances clarification due to the flocculant nature of the $Mg(OH)_2$. Excess calcium ions at high pH levels may be precipitated by the addition of soda ash. The following reactions are summarized:

$$Ca(OH)_2 + Ca(HCO_3)_2 \rightarrow 2CaCO_3\downarrow + 2H_2O \qquad (8-13)$$

$$2Ca(OH)_2 + Mg(HCO_3)_2 \rightarrow$$
$$2CaCO_3\downarrow + Mg(OH)_2\downarrow + 2H_2O \quad (8-14)$$

$$Ca(OH)_2 + Na_2CO_3 \rightarrow CaCO_3\downarrow + 2NaOH \qquad (8-15)$$

Reduction of high pH levels is accomplished as follows: A two-stage method employs first the precipitation of calcium carbonate through the addition of carbon dioxide:

$$Ca(OH)_2 + CO_2 \rightarrow CaCO_3\downarrow + H_2O \qquad (8-16)$$

Single-stage pH reduction is accomplished by carbon dioxide addition (or acids). This reaction is the second stage of the two-stage method:

$$Ca(OH)_2 + 2CO_2 \rightarrow Ca(HCO_3)_2 \qquad (8-17)$$

The lime demand of a given wastewater is a function of the buffer capacity or alkalinity of the wastewater.

Soda ash (Na_2CO_3) is available in two forms: light soda (bulk density range of 35 to 50 lb/ft^3) and a dense soda ash (density range of 60 to 76 lb/ft^3). The pH of a one percent solution of soda ash is 11.2. It is used for pH control and in lime treatment.

The molecular weight of soda ash is 106. Commercial purity ranges from 98 to greater than 99% Na_2CO_3. The viscosities of sodium carbonate solutions are given in Figure 8-6. Soda ash in the presence of lime and water forms caustic soda.

Liquid caustic soda is commercially available at two concentrations, 50% and 73% NaOH. The densities of the solutions as shipped are 12.76 lb/gal for the 50% solution and 14.18 lb/gal for the 73% solution. Solutions contain 6.38 lb/gal NaOH and 10.34 lb/gal NaOH,

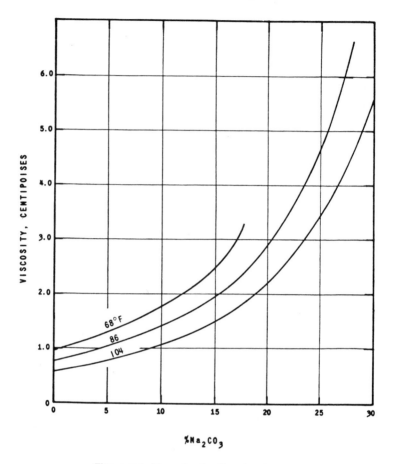

Figure 8-6. Viscosity of soda ash solutions.

respectively. The crystallization temperature is 53°F for the 50% solution and 165°F for the 73% solution. Molecular weight is 40. Viscosities of various caustic soda solutions are presented in Figure 8-7. The pH of a one percent solution of caustic soda is 12.9.

Carbon dioxide (CO_2) is available in gas and liquid form. Molecular weight is 44. Dry CO_2 is not chemically active at normal temperatures and is a nontoxic chemical. The gas displaces oxygen and adequate ventilation of closed areas should be provided. Solutions of CO_2 in water are very reactive, forming carbonic acid. Saturated solutions of CO_2 have a pH of 4.0 at 68°F.

The gas form may be produced on the treatment plant site by scrubbing and compressing the combustion product of lime recalcining furnaces, sludge furnaces, or generators used principally for the production of CO_2 gas only. These generators are usually fired with combustible gases, fuel oil, or coke. The plant site gas forms usually have a CO_2 content of between 6 and 18% depending on the source and efficiency of the producing system. The commercial liquid form has a minimum CO_2 content of 99.5%.

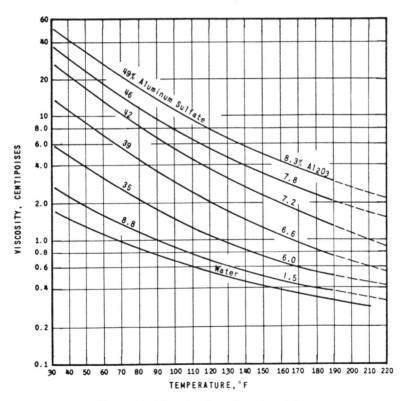

Figure 8-7. Viscosity of caustic soda solutions.

Notes on Chemical Feeders

The capacity of a chemical feed system is an important consideration in both storage and feeding. Storage tanks or bins for solid chemicals must be designed with proper consideration of the angle of repose of solid material and environmental requirements, such as temperature and humidity. Size and slope of feeding lines are important along with construction materials with respect to corrosiveness.

Feeders must accommodate minimum and maximum feeding rates. Manually controlled feeders have a common range of 20:1, but this range can be increased to about 100:1 with dual control systems. Chemical feeder control can be manual, automatically proportioned to flow, dependent on some form of process feedback or a combination of any two of these. Sophisticated control systems are also available. If manual control systems are specified with the possibility of future automation, the feeders selected should be amenable to this conversion with a minimum of expense. Standby or backup units should be included for each type of feeder used. Reliability calculations will be necessary in larger plants with a greater multiplicity of units. Points of chemical addition and piping to them should be capable of handling all possible changes in dosing patterns in order to have flexibility. Designed flexibility in hoppers, tanks, chemical feeders, and solution lines is important to maximum benefits at minimal cost.

Liquid feeders are generally of metering pumps or orifices. Metering pumps are of the positive-displacement variety, plunger or diaphragm type. The choice of liquid feeder is highly dependent on the viscosity, corrosivity, solubility, suction and discharge heads, and internal pressure-relief requirements. Examples are shown in Figure 8-8. In some cases control valves and rotameters may be all that is required. In other cases, such as lime slurry feeding, centrifugal pumps with open impellers are used with appropriate controls. More complete descriptions of liquid feeder requirements can be found in References 1, 2, and 3.

Solids characteristics vary greatly and feeder selection must be carefully considered. Provisions should be made to keep all chemicals cool and dry. Hygroscopic (water absorbing) chemicals may become lumpy, viscous or even rock hard; other chemicals with less affinity for water may become sticky from moisture on the particulate surfaces. This causes increased arching or bridging in hoppers. Moisture affects the density of the chemial and may result in under-feed.

Volumetric feeding of solids is normally restricted to smaller plants, specific types of chemicals which are reliably constant in composition and low rates of feed. Several volumetric types are available. Accuracy

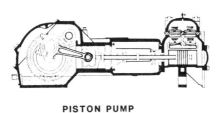

PISTON PUMP

STROKE ACTUATOR

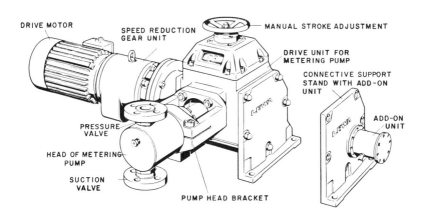

DRIVE MOTOR

SPEED REDUCTION GEAR UNIT

MANUAL STROKE ADJUSTMENT

DRIVE UNIT FOR METERING PUMP

CONNECTIVE SUPPORT STAND WITH ADD-ON UNIT

ADD-ON UNIT

PRESSURE VALVE

HEAD OF METERING PUMP

SUCTION VALVE

PUMP HEAD BRACKET

DIAPHRAGM PUMP

Figure 8-8. Positive displacement pumps.

of feed is usually limited to $\pm 2\%$ by weight but may be as high as $\pm 15\%$.

One common type of volumetric dry feeder uses a continuous belt to transfer material to the dissolving tank. A mechanical gate regulates the depth of material on the belt, and the rate of feed is governed by the speed of the belt and/or the height of the gate opening. The hopper normally is equipped with a vibratory mechanism to reduce solids arching. This type of feeder is not suited for easily fluidized materials.

Another type uses a screw or helix from the bottom of the hopper through a tube opening slightly larger than the diameter of the screw or helix. Feed rate is controlled by the screw or helix rotation speed. Some screw-type designs are self-cleaning, while others clog. Figure 8-9 shows a typical screw-feeder.

Other volumetric feeders fall into the positive-displacement class. Designs of this type employ some form of moving cavity of a specific or variable size. In operation, the chemical falls by gravity into the cav-

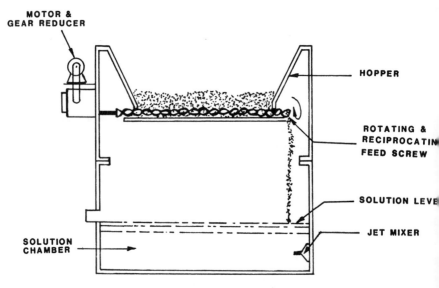

Figure 8-9. Screw feeder.

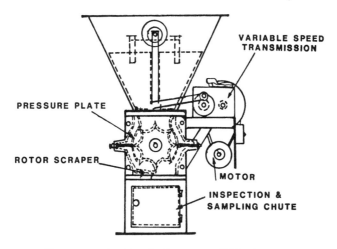

Figure 8-10. Positive displacement solid feeder-rotary.

ity and is more or less fully enclosed and separated from the hopper's feed. The size of the cavity, and the rate at which the cavity moves and discharges, governs the feed rate. The positive control of the chemical may place a low limit on rates of feed. The progressive cavity metering pump is a nonreciprocating type. Positive-displacement feeders often utilize air injection to improve the flow of the material. One example of positive-displacement units is shown in Figure 8-10.

Volumetric solids feeders can compensate for changes in material density, by including a gravimetric controller. This modification allows for weighing of the material as it is fed. A beam balance type measures the actual mass of material. This is more accurate, particularly over a long period of time, than the less common spring-loaded gravimetric designs. Gravimetric feeders are used where feed accuracy of about one percent is required, as in large-scale operations and for materials which are used in small, precise quantities. It should be noted, however, that even gravimetric feeders cannot compensate for weight added to the chemical by excess moisture.

Table 8-1 summarizes major types of feeders and their characteristics.

Table 8-1
Types of Chemical Feeders

Type of Feeder	Feed Materials	Capacity ft³/hr	Range
Dry feeder:			
Volumetric:			
Oscillating plate	Any material, granules or powder.	0.01 to 35	40 to 1
Oscillating throat (universal)	Any material, any particle size.	0.02 to 100	40 to 1
Rotating disc	Most materials including NaF, granules or powder.	0.01 to 1.0	20 to 1
Rotating cylinder (star)	Any material, granules or powder.	8 to 2,000 or 7.2 to 300	10 to 1 or 100 to 1
Screw	Dry, free flowing material, powder or granular.	0.05 to 18	20 to 1
Ribbon	Dry, free flowing material, powder, granular, or lumps.	0.002 to 0.16	10 to 1
Belt	Dry, free flowing material up to 1½-inch size, powder or granular.	0.1 to 3,000	10 to 1 or 100 to 1
Gravimetric:			
Continuous—belt and scale	Dry, free flowing, granular material, or floodable material.	0.02 to 2	100 to 1
Loss in weight	Most materials, powder, granular or lumps.	0.02 to 80	100 to 1
Solution feeder:			
Nonpositive displacement:			
Decanter (lowering pipe) . . .	Most solutions or light slurries	0.01 to 10	100 to 1
Orifice	Most solutions	0.16 to 5	10 to 1
Rotameter (calibrated valve)	Clear solutions	0.005 to 0.16 or 0.01 to 20	10 to 1
Loss in weight (tank with control valve).	Most solutions	0.002 to 0.20	30 to 1
Positive displacement:			
Rotating dipper	Most solutions or slurries	0.1 to 30	100 to 1
Proportioning pump:			
Diaphragm	Most solutions. Special unit for 5% slurries.[1]	0.004 to 0.15	100 to 1
Piston	Most solutions, light slurries. .	0.01 to 170	20 to 1
Gas feeders:			
Solution feed	Chlorine	8000 lb/day max	20 to 1
	Ammonia	2000 lb/day max	20 to 1
	Sulfur dioxide	7600 lb/day max	20 to 1
	Carbon dioxide	6000 lb/day max	20 to 1
Direct feed	Chlorine	300 lb/day max	10 to 1
	Ammonia	120 lb/day max	7 to 1
	Carbon dioxide	10,000 lb/day max	20 to 1

References

1. Cheremisinoff, N. P., *Applied Fluid Flow Measurement,* Marcel Dekker Inc., New York (1979).
2. Cheremisinoff, N. P., *Fluid Flow: Pumps, Pipes and Channels,* Ann Arbor Science Pub., Ann Arbor, MI (1981).
3. Cheremisinoff, P. N. and R. A. Young, *Pollution Engineering Practice Handbook,* Ann Arbor Science Pub., Ann Arbor, MI (1976).

INDEX